Sustainable Food System Assessment

Sustainable Food System Assessment provides both practical and theoretical insights about the growing interest in and response to measuring food system sustainability. Bringing together research from the Global North and South, this book shares lessons learned, explores intended and actual project outcomes, and highlights points of conceptual and methodological convergence.

Interest in assessing food system sustainability is growing, as evidenced by the Milan Urban Food Policy Pact and the importance food systems initiatives have taken in serving as a lever for attaining the UN Sustainable Development Goals. This book opens by looking at the conceptual considerations of food systems indicators, including the place-based dimensions of food systems indicators and how measurements are implicated in sense-making and visioning processes. Chapters in the second part cover operationalizing metrics, including the development of food systems indicator frameworks, degrees of indicator complexities, and practical constraints to assessment. The final part focuses on the outcomes of assessment projects, including impacts on food policy and communities involved, highlighting the importance of building connections between sustainable food systems initiatives.

The global coverage and multi-scalar perspectives, including both conceptual and practical aspects, make this a key resource for academics and practitioners across planning, geography, urban studies, food studies, and research methods. It will also be of interest to government officials and those working within NGOs.

Alison Blay-Palmer is Director, Centre for Sustainable Food Systems and Professor of Geography and Environmental Studies, Wilfrid Laurier University, Canada.

Damien Conaré is Secretary General of the UNESCO Chair in World Food Systems, Montpellier SupAgro, France.

Ken Meter is President of the Crossroads Resource Center, USA.

Amanda Di Battista is a Project Coordinator at the Laurier Centre for Sustainable Food Systems, Canada.

Carla Johnston is a PhD candidate, Balsillie School of International Affairs, Laurier Center for Sustainable Food Systems, Canada.

Routledge Studies in Food, Society and the Environment

Urban Food Systems Governance and Poverty in African Cities
Edited by Jane Battersby and Vanessa Watson

Risk and Food Safety in China and Japan
Theoretical Perspectives and Empirical Insights
Edited by Louis Augustin-Jean and Jean-Pierre Poulain

Localizing Global Food
Short food supply chains as responses to Agri-Food System challenges
Edited by Sophia Skordili and Agni Kalfagianni

Seafood Supply Chains
Governance, Power and Regulation
Miriam Greenwood

Civil Society and Social Movements in Food System Governance
Edited by Peter Andrée, Jill K. Clark, Charles Z. Levkoe and Kristen Lowitt

Voice and Participation in Global Food Politics
Alana Mann

Plant-Based Diets for Succulence and Sustainability
Edited by Kathleen May Kevany

Sustainable Food System Assessment
Lessons from Global Practice
Edited by Alison Blay-Palmer, Damien Conaré, Ken Meter, Amanda Di Battista, and Carla Johnston

For further details please visit the series page on the Routledge website: http://www.routledge.com/books/series/RSFSE/

Sustainable Food System Assessment

Lessons from Global Practice

Edited by Alison Blay-Palmer, Damien Conaré, Ken Meter, Amanda Di Battista, and Carla Johnston

Routledge
Taylor & Francis Group
LONDON AND NEW YORK

earthscan
from Routledge

First published 2020
by Routledge
2 Park Square, Milton Park, Abingdon, Oxon OX14 4RN

and by Routledge
52 Vanderbilt Avenue, New York, NY 10017

Routledge is an imprint of the Taylor & Francis Group, an informa business

First issued in paperback 2021

British Library Cataloguing-in-Publication Data
A catalogue record for this book is available from the British Library

Library of Congress Cataloging-in-Publication Data
Names: Blay-Palmer, Alison, 1961– editor. |
Conaré, Damien, editor. | Meter, Ken, editor. |
Di Battista, Amanda, editor. | Johnston, Carla, editor.
Title: Sustainable food system assessment : lessons from
global practice / edited by Alison Blay-Palmer, Damien Conaré,
Ken Meter, Amanda Di Battista, Carla Johnston.
Description: Milton Park, Abingdon, Oxon ;
New York, NY : Routledge, 2020. |
Series: Routledge studies in food, society and the environment |
Includes bibliographical references and index.
Identifiers: LCCN 2019031769 (print) | LCCN 2019031770 (ebook) |
ISBN 9781138341951 (hardback) | ISBN 9780429439896 (ebook)
Subjects: LCSH: Food supply–Environmental aspects. |
Sustainable agriculture. | Nutrition policy.
Classification: LCC HD9000.5 .S83395 2020 (print) |
LCC HD9000.5 (ebook) | DDC 338.1/9–dc23
LC record available at https://lccn.loc.gov/2019031769
LC ebook record available at https://lccn.loc.gov/2019031770

ISBN: 978-1-138-34195-1 (hbk)
ISBN: 978-1-03-208393-3 (pbk)
ISBN: 978-0-429-43989-6 (ebk)

Typeset in Sabon
by Newgen Publishing UK

To the memory of Marielle Dubbeling, RUAF Foundation co-founder, leading international expert in urban agriculture and city region food systems, researcher, teacher and visionary. 1968–2019

Contents

Figures

Tables

Contributors

Thomas Allen is a research economist at the Banque de France specialized in green finance. He holds a PhD in Economics from the University of Montpellier. He has previously worked at the OECD and at Bioversity International (CGIAR) on sustainable food systems.

Gabriel April-Lalonde acquired a solid foundation in human nutrition during his Bachelor's degree, which he continues to develop and strengthen through academic and professional projects. His Master's degree in Nutrition allowed him to explore diet in relation to public health, environment, and education. He is now pursuing a PhD in Nutrition at the Université de Montréal, to investigate how more or less socially and environmentally responsible food behaviours are associated with human health.

Jennifer Baltzer is a Canada Research Chair in Forests and Global Change at Wilfrid Laurier University (Laurier), whose work focuses on the drivers of forest composition, structure, and function and responses of these systems to global change. Within Canada, she works closely with the Government of the Northwest Territories (GNWT) through a 10-year Partnership Agreement between the GNWT and Laurier, and internationally is involved in several key collaborations through NASA and the Smithsonian Institute. She plays a leadership role in the CFREF-funded Global Water Futures program and was elected to the Royal Society of Canada's College of New Scholars, Artists, and Scientists in 2017.

Jane Battersby is a senior researcher at the African Centre for Cities, University of Cape Town, South Africa, and is PI of the IDRC-funded Nourishing Spaces Project. An urban geographer by training, her work focuses on urban food security, food systems, and governance.

Peter R. Berti joined HealthBridge in 1998, where he is Deputy Director and Nutrition Advisor. He is responsible for ensuring the technical quality of all of HealthBridge's work in food and nutrition research and programming. Much of his work has been in the Americas (Ecuador, Bolivia, and Brazil) where he has conducted research and managed programmes in food system research, nutritional assessment, and food fortification.

Alison Blay-Palmer is a Professor in the Department of Geography and Environmental Studies and the Director of the Laurier Centre for Sustainable Food Systems at the Balsillie School for International Affairs at Wilfrid Laurier University in Waterloo, ON Canada.

Laura Boada has a Bachelor's degree in Food Science and Technology with a Master's degree in Rural Development. Her research focus is on food systems in rural and urban environments. She has designed and developed a food process pilot plant in Quito, Ecuador.

Veronica Bonomelli is a PhD student in Social Sciences at Université Libre de Bruxelles (Belgium). Her research interests focus on food-related issues, with emphasis on urban food policies and their articulations with non-public initiatives claiming more sustainable food systems.

Nicolas Bricas is a scientist in the socio-economics of food systems at Cirad, the French agricultural research and international cooperation organization working for the sustainable development of tropical and Mediterranean regions. His main research topics are the effects of rapid urbanization, industrialization, and liberalization on food systems in African and Asian countries. He is currently the Director of the Unesco Chair on World Food Systems, at Montpellier SupAgro and Cirad, where he leads research programmes and training sessions on the Sustainable Urban Food System, in France and abroad.

Bruce Cogill is a research leader, global coordinator, project director, and a food, health, and nutrition security specialist and advisor. He has over 30 years of emergency and development-related experience with the UN, governments, universities, and foundations, mostly in Sub-Saharan Africa, South Asia, and the South Pacific. He recently directed a global team of trans-disciplinary researchers at the Rome-based CGIAR centre, Bioversity International, studying agricultural biodiversity, dietary diversity, sustainable food systems/diets, health, and nutrition. He is a consultant and advisor to the Department of Foreign Affairs (DFAT), Government of Australia, a technical advisor for the Power of Nutrition, the Wellcome Trust Foundation, universities, and technical groups. He is Associate Editor for nutrition for the journal *Global Health: Science and Practice*. He holds a PhD in nutrition and agricultural economics from Cornell University. In addition to scientific publications, he has presented at numerous venues and lectured at universities, including the Mailman School of Public Health and Columbia University, for over 15 years.

Nevin Cohen is an Associate Professor at the City University of New York (CUNY) Graduate School of Public Health and Health Policy, and is Research Director of the CUNY Urban Food Policy Institute. His research explores the policies, governance systems, practices, and infrastructure to support socially just, healthy, ecologically resilient, and economically viable urban and regional food systems. He has a PhD in Urban Planning and Policy Development from Rutgers University, a Master's in City and Regional Planning from Berkeley, and a BA from Cornell.

Donald C. Cole is a public, occupational, and environmental health physician, with a Master's degree in health research methods, and post-doctoral studies in ecosystem approaches to health. During over 35 years of practice (clinical and population-related), research and policy work in Canada and lower- and middle-income countries, he has emphasized multi-stakeholder action research processes to change socio-ecological conditions to improve physical and mental health. He currently works part-time as a clinician and part-time in teaching, mentoring, and conducting research as a professor at the University of Toronto's Dalla Lana School of Public Health.

Damien Conaré is the Secretary General of the UNESCO Chair in World Food Systems, hosted by Montpellier SupAgro (the French National Institute of Higher Education in Agricultural Sciences). He was previously editor-in-chief of the quarterly journal *Courrier de la planète*, edited by the NGO Solagral.

Amanda Di Battista is the Project Coordinator for the Laurier Centre for Sustainable Food Systems at Wilfrid Laurier University in Waterloo, Ontario. She has a Master's degree in Environmental Studies with a research focus on critical environmental pedagogies. She is co-editor of *The Goose* and produces two academic podcasts, *Handpicked: Stories from the Field* and *CoHearence*.

Marielle Dubbeling is director of the RUAF Foundation, she has over 20 years' experience in Urban Agriculture and Sustainable Urban Food Systems, she has working experience in Asia, Africa, Europe, and Latin America and has authored several publications.

Olivier Lepiller, sociologist PhD, permanent researcher at Cirad (MOISA unit, Montpellier, France), is responsible for a work package in the URBAL project. He specializes in the sociology of food and health and has published papers on topics such as the criticisms of industrial food, the consumption of animal-source foods, the sociocultural forms of food transitions, the social making of food heritage, and the consumption of dietary supplements.

Ken Meter, president of Crossroads Resource Center in Minneapolis, is one of the most experienced food system analysts in the US. His work integrates market analysis, business development, systems thinking, and social concerns. Meter holds 49 years of experience in inner-city and rural community capacity building. He developed a $9.85 million plan for local food investment for the state of South Carolina, and completed statewide food system assessments for Hawai'i, Alaska, Mississippi, Indiana, Ohio, and Minnesota.

Ana Moragues-Faus is a Research Fellow at the Sustainable Places Institute, in Cardiff University. She combines a set of critical theories such as political ecology and participative justice to support the development of more just and sustainable food systems. Her research interests revolve mainly around sustainable agri-food systems, governance, and food security, having developed most of her work in European countries and Latin America. At the moment she is researching, through participative processes, the emergence and unfolding of urban food governance initiatives across the globe and exploring their role in transforming the food system.

Fabián Muñoz has Master's degrees in Communication (with a focus on public opinion surveys) and Visual Analytics and Big Data. He is a statistical consultant specializing in population sampling, survey methods, multivariable analysis and communication, and visualization of data with various health and nutritional research collaborations. He also teaches applied statistics at the university level and is manager of the consulting company Visor Análisis Estadística Cía. Ltda.

Erin Nelson's research explores the development of more sustainable food systems with a particular focus on agroecology initiatives in both Canada and Latin America. In particular, she looks at how knowledge-exchange can build capacity for agroecological production, and how agroecology can contribute to ecological resilience and community well-being. She works in close collaboration with a wide range of partners, including civil society organizations and farmer networks.

Anne Palmer is the Food Communities and Public Health programme director at the Johns Hopkins Center for a Livable Future and a senior research associate at the Johns Hopkins Bloomberg School of Public Health in the Department of Health, Behavior, and Society. She directs the Food Policy Networks project, which seeks to improve the capacity of food policy councils and similar organizations to advance food system policies.

Myriam Paredes is the academic coordinator of the department of Development, Environment, and Territory at FLACSO, Ecuador. Her work is to ensure the quality of research, teaching, and communication of Master's and doctoral programmes. Her research focuses on heterogeneity studies to understand the different faces of Food Sovereignty.

Priscila Prado Beltrán is a PhD student in the Agroecosystem Program of Federal University of Santa Catarina-Brasil. She has worked with rural communities and NGOs developing sustainable research projects related to agroecology, health, solidarity economy, and food. Her research thesis is part of the Ekomer project.

Paolo Prosperi is a researcher in agricultural and food economics at the University of Pisa (Department of Agriculture, Food and Environment) where he teaches Food Policy, Agricultural Economics, and Natural Resource Management. He holds a PhD in Economics from the University of Montpellier and his research fields include assessment methods for sustainable food systems and food security, the adaptive strategies of agriculture and fisheries producers, as well as small farming dynamics. He is involved in a number of European transdisciplinary projects on food systems' sustainability (SUFISA, SALSA, LIAISON, DESIRA).

Guido Santini leads the FAO City Region Food Systems Programme that provides support for cities and their regions to build more sustainable and resilient food systems and to strengthen rural–urban linkages. Guido has an MSc degree in tropical and subtropical agriculture (awarded by University of Florence, Italy and the Technical University of Lisbon, Portugal). He has 15 years' work experience in the field of international development cooperation within the United Nations system (FAO and UNDP) and research institutions. This encompasses a range of programme/project design and management, policy support and research activities in the areas of natural resources management (land and water), agricultural, rural and urban development, and food security, with experience in countries in Europe, Sub-Saharan Africa, Asia, and Latin America. In particular, Guido has conducted long-term assignments in Europe (Albania) and Sub-Saharan Africa (Mozambique, Kenya, and Uganda).

Denis Sautier is a researcher in agrifood economics at CIRAD, member of the "Innovation" research unit. He works on localized food systems and land-based food labelling in Asia, Africa and Latin America and for six years was the facilitator of the Hanoi-based research platform "Markets and Agriculture Linkages for Cities in Asia" (MALICA).

Raychel Santo is a Senior Research Program Coordinator at the Johns Hopkins Center for a Livable Future, where she works on research projects related to local and regional food governance, urban agriculture, institutional food procurement, and the relationship between diet and climate change. Raychel earned her Master's degree in Food, Space & Society from Cardiff University School of Geography & Planning and her BA in Public Health and Environmental Change & Sustainability from Johns Hopkins University.

Kerstin Schreiber is a Doctoral Student in Geography at McGill University, Canada, working in the Land & Food Laboratory of Prof. Graham MacDonald. She is a scholar in the Sustaining Landscapes Graduate Student Program of the McGill Sustainability Systems Initiative (MSSI) and is lead organizer of the affiliated annual Sustainability Research Symposium for students. Her research interests include urban food systems, food supply chains and the social, ecological, and economic factors that link cities and their agricultural hinterlands.

Melaine Simba is the Environmental Coordinator for the Ka'a'gee Tu First Nation and Technical Advisor for the Aboriginal Aquatic Resource and Oceans Management (AAROM) program for the Dehcho First Nation.

Kelly Skinner is an Assistant Professor and Program Lead, Master of Health Evaluation at the School of Public Health and Health Systems at the University of Waterloo, Waterloo, Ontario, Canada. The primary focus of her research has been community-based health and social projects related to food, nutrition, food security, health and risk communication of contaminants in traditional food sources, and the broader context of food systems and environments in northern Canada.

Andrew Spring is the Research Associate in Northern Knowledge Networks, Adjunct Professor in the Department of Geography and Environmental Studies, and Associate Director, Laurier Centre for Sustainable Food Systems at Wilfrid Laurier University, Waterloo, ON.

Heidi Swanson is Assistant Professor and University Research Chair in the Department of Biology and the University of Waterloo, Waterloo ON.

Merritt Turetsky is Associate Professor and Canada Research Chair, Tier 2 in the Department of at the University of Guelph, Guelph, ON.

Yubari Valero is a PhD candidate in Development Economics at FLACSO-Ecuador. She has an MSc in Social Economy from General Sarmiento University – Argentina, along with two Specializations, one in Local Economic Development from the Center for Rural Urban Studies (CERUR), Israel and another in Endogenous Development of the Rural Sector, from the Universidad Centroccidental Lisandro Alvarado (UCLA), Venezuela. Her research focuses on rural development, food sovereignty, and social and solidary economies.

Elodie Valette is a geographer and a senior researcher at CIRAD (French Agricultural Research for Development). Her research focuses on territorial approaches for development, especially addressing the rural–urban linkages in France and in the Mediterranean. She is one of the coordinators of the URBAL research project, focusing on characterizing the impact of urban-driven food innovations on the sustainability of foods systems.

1 Sustainable food system assessment

Lessons from global practice

Alison Blay-Palmer, Damien Conaré, Ken Meter, and Amanda Di Battista

Introduction

Sustainable Food System Assessment: Lessons from Global Practice has its roots in workshops with participants in the Food: Locally Embedded, Globally Engaged (FLEdGE) Partnership as part of the work by the Laurier Centre for Sustainable Food Systems (LCSFS). The first workshop took place in Toronto, Canada in June 2016 with a follow-up meeting in Waterloo in September 2017. The goal of the gatherings was to share lessons learned and develop collective insights for moving forward with the various projects and the expert participants exploring opportunities for comparative work. These meetings provided a rich starting point, given the broad inclusion of academics and practitioners working on assessment projects developed from the city region to the national scale in both the Global South and North. Central observations, considerations, and questions from these meetings can be grouped as: (1) conceptual considerations including sense-making, vision, and place; (2) operationalization of metrics specifically developing frameworks and representations, determining degrees of complexity, the challenges of working within and across scales, and dealing with practical constraints such as data availability; and (3) outcomes and goals for assessment projects including policy generation, community inclusion and participation, building connections between initiatives, embedding change in communities, and knowledge dissemination. Taken together, these three overlapping themes capture the process of developing sustainable food systems assessment (SFSA) approaches from vision and concept, through operationalization, to ending with outputs (and in some cases, impacts). As such, these themes provide the framework for our book.

We begin with a scan of existing literature to provide context. Following, we provide a review of the workshop conclusions, including references to selected relevant literatures. Finally, this chapter includes a discussion of the chapters in this book. We revisit the core concepts in Chapter 12.

Insights from the sustainable food system assessment literature: terms of reference, context, and assessment considerations

Exploring terms and meaning

Recognizing the tensions around the words 'sustainability' and 'systems', it is useful to bracket how we use 'sustainable food systems' (SFS) in this book. As Prosperi et al. in Chapter 7 explain from their research in trying to understand sustainability, 'People want a descriptor of a state rather than the prediction of a state'. With this in mind, we provide specific criteria for describing what constitutes an SFS.

While acknowledging that sustainability is a contested term, for our purposes 'sustainable' builds on the three-pillar approach from *Our Common Future* (Brundtland et al., 1986), used by many including the United Nation's *2030 Agenda for Sustainable Development* and the associated Sustainable Development Goals (SDGs). Its three overlapping considerations are social, environmental, and economic. In a food systems context, social dimensions include the Right to Food, and ensuring food and nutrition security, food democracy, fair labour practices, gender equity, social connectivity, cultural self-determination, and natural resource rights including secure land tenure. Environmental considerations include ecological food production methods that acknowledge the important role of agroecology, biodiversity, renewable energy sources, and protecting the quality of soils, water, and other resources, while working towards regenerative closed loop food systems. Economic dimensions build from the premise of keeping equitable economic activity at the local as much as possible and then moving outward. This fosters supportive, circular commercial networks and infrastructure that include developing mutual trust and equal sharing of value and risk across agro-food networks from local to global. This is an important consideration as localization alone does not guarantee fair economic relations (Born & Purcell, 2006). The goal is to enhance community economic development through short, alternative food networks with models that include co-operatives, community supported agriculture and other forward investments, food sharing, collaborative business networks, and social economy approaches. Finally, inclusive, transparent, participatory, and democratic governance mechanisms are critical to support the three sustainability dimensions and are foundational to their success (Feenstra, 1997; Bricas, 2017; Blay-Palmer et al., 2018).

Given the complex, diverse, and necessarily adaptive demands of working towards sustainability, this book draws upon systems lenses to understand the possibilities for bringing about transformation through food collaborations (Stroink & Nelson, 2013; Knezevic et al., 2017; Chapter 4, this volume[1]). While these systems lenses derive from many sources (Hipel et al., 2010; Ingram, 2011; Blay-Palmer et al., 2015; Hinrichs, 2016; Meter, 2007), we build explicitly from *Thinking in Systems* where Meadows (2008) defines a system as, 'an interconnected set of elements that is coherently organized in a way that achieves something' (p. 12). Systems are 'more than the sum of their parts' and can be '... adaptive, dynamic, goal-seeking, self-preserving, and sometimes, evolutionary' (p. 12). While there is integrity to systems and mechanisms to maintain balance, systems are also able to '... be self-organising, and are often self-repairing at least over some range of disruptions' (p. 12). Considering the myriad implications at the intersection of the definitions for sustainability and systems, it becomes clear that developing assessment tools and processes can be challenging (Stroink & Nelson, 2013; Chapters 4 & 7, this volume). That said, there are many complex and useful approaches to sustainable food systems that inform this book.

Next, we review some of the broader context that has fostered the emergence of assessment as part of the way forward for sustainable food systems. Academic work has increasingly embraced the perspectives and work of hundreds of community-based initiatives, amplifying the efforts of grassroots projects while codifying lessons that can be applied across contexts.

The emergence of sustainable food system assessments

The International Assessment of Agricultural Knowledge, Science and Technology for Development (IAASTD) is a key starting point for understanding current approaches to food system sustainability indicators as it helped to frame, both directly and indirectly, how actors consider and work within the existing food system at multiple scales. The IAASTD consultation and subsequent reports resulted from a multi-year (2004–2008) and multi-stakeholder process that included a wide spectrum of experts from research institutions and civil society, including both public and private sectors. The IAASTD process was developed to inform policy formulation around research and knowledge creation for SFS using agriculture as the starting point. It explicitly pushed back against the dominant assumption about high technology, scientific interventions alone and valued the knowledge and experiences of traditional, smallholder farmers and consumers. A primary goal was to present a multi-sectoral and integrated review from multiple world views so that,

> the IAASTD does not advocate specific policies or practices ... It is policy relevant, but not policy prescriptive. It integrates scientific information on a range of topics that are critically interlinked, but often addressed independently, i.e., agriculture, poverty, hunger, human health, natural resources, environment, development and innovation. It will enable decision makers to bring a richer base of knowledge to bear on policy and management decisions on issues previously viewed in isolation ... presents different views, acknowledging that there can be more than one interpretation of the same evidence based on different worldviews; and identifies the key scientific uncertainties and

> areas on which research could be focused to advance development and sustainability goals.
>
> (IAASTD, 2009, pp. vii–viii)

In this way IAASTD was part of a watershed moment in opening-up the consultation process to include smallholder farmers' knowledge using agroecological and other traditional practices. Other critical and formative events unfolded as the final pages of IAASTD were written: the reform of the United Nations Committee on World Food Security (CFS) in 2009 in the wake of the 2008–2009 food crisis (Anderson, 2015; McKeon, 2015), as well as the launch by the FAO of a consultation process to develop its *Voluntary Guidelines on the Responsible Governance of Tenure of Land, Fisheries and Forests in the Context of National Food Security* (published in 2012). There was also a revival of a focus on the Right to Adequate Food and Nutrition (FIAN, 2016), as well as the increasing role of La Via Campesina, and the 2009 *People's Food Sovereignty Now!* declaration by the Civil Society Organization (CSO) Forum, which ran parallel to the World Summit on Food Security in Rome. These clarified that civil society needs to be a key contributor in moving the sustainable food systems agenda forward.

The International Panel of Experts on Sustainable Food Systems (IPES-Food) 2015 report reiterated concerns expressed in IAASTD around power and the political economy of knowledge that result in disjointed and siloed approaches to, and identification of, sustainability solutions. To overcome this challenge and develop more coherent approaches to food system sustainability, the IPES-Food analytical framework called for analysing

> Webs of complex interactions and feedback loops in food systems; broad constellations of policies with the capacity to affect food systems; power relations and the political economy of food systems; a multi-scale and holistic understanding of sustainability, as the benchmark of food systems reform.
>
> (IPES-Food, 2015, p. 3)

It also put forward the need to foster a new transdisciplinary science of food systems, one that requires scholars to break down boundaries and silos between disciplines and around knowledge, encouraging the co-creation of knowledge with civil society (IPES-Food, 2015, p. 8).

In addition to recognizing the importance of traditional food system knowledge and the interconnectedness of food systems, 'measuring' change emerged as a priority for understanding more about SFS. As a result, indicators gained importance at all scales for policymakers, researchers, and funders, with metrics seen as the way to benchmark, assess, and track food system sustainability from cities to the global scale.

Recent key examples that demonstrate movement in this direction range from the urban–regional-focused *Milan Urban Food Policy Pact* (MUFPP) to the globally scaled SDGs. At the municipal and regional scale, the MUFPP uses six categories to understand and foster food system sustainability. These include effective governance, sustainable diets and nutrition, social and economic equity, food production, food supply and distribution, and waste. With more than 200 signatory cities, the MUFPP is enabling food system sustainability at the city–region scale. Since 2017, the MUFPP offers 44 indicators with four to ten indicators per category (Calori et al., 2017).

Assessment tools are also well established at the sub-national and more local contexts, for example the *Calgary Food Action Plan – Calgary Eats!*, the *Vancouver Food System Assessment*, and the *Toronto Food Strategy*. Urban metric-based assessments have also been undertaken for specific parts of an SFS and can enable comparative analysis. For example, work in Cape Town, South Africa, drawing on individual and household food security survey data, reported on the links between household food insecurity, income, and informal food sector markets and informal social safety nets. The analysis identified that the lower a person's or household's income, the more likely they were to rely on informal networks to secure their food (Battersby, 2011). Data were gathered through an 11-city project in southern Africa and allowed for some comparisons across cities (e.g. Crush et al., 2012). Another example is the work of the Sustainable Food Cities project in the UK and its report *Urban Food Strategies: The Rough Guide to Sustainable Food Systems* (Moragues et al., 2013; Chapter 6, this volume) that connects the realities of the local food system to the broader global scale, providing insights into how communities can use food systems initiatives to counter global pressures. By identifying several community well-being factors, including health, environmental impacts, economic performance, injustice, and cultural erosion, the assessment demonstrated that urban food strategies are locally contingent, and that local engagement varies. The research further showed the need for local engagement by key actors (Moragues et al., 2013, p. 6).

The City Region Food Systems (CRFS) project documented place-specific sustainability dimensions of food flows for key local staple foods in both the Global South and North. This work has enabled multi-scaled, multi-actor policy initiatives and networks with a view to improving various dimensions of the food system, including urban–rural linkages, food access (especially for low-income families), waste management and improved incomes for rural and urban producers (Dubbeling et al., 2017; Blay-Palmer et al., 2018; Chapter 9, this volume). At the more micro scale, the UN Food Insecurity Experience Scale (FIES) provides a snapshot of household food insecurity and can act as a rapid assessment tool for practitioners, complementing other tools that encompass the probability of undernourishment and measures of food insecurity determinants. At the farm scale, the Response Inducing Sustainability Evaluation (RISE) relies on interviews and then computer evaluation to score farm-level sustainability (Grenz et al., 2011) while research on Flemish dairy farms (MOTIFS) was designed to measure integrated farm sustainability using ecological, economic, and social themes to identify indicators that could be gathered simply (Meul et al., 2008).

There are several tools that assess dimensions of sustainable food systems within countries. For example, the Sustainability Assessment of Food and Agriculture (SAFA) provides guidance for national-level assessment. SAFA was created to apply universal sustainability goals to food value chains. It was developed to be holistic, addressing all dimensions of sustainability (including environmental, social, economic, and governance) and applicable to all operational scales. *Food Counts: the pan-Canadian Sustainable Food Systems Report Card*, provides existing metrics and identifies information gaps across several food sovereignty pillars, named as: provides food for people, values providers, works with nature, localizes food, puts control locally, and puts food as sacred (Levkoe & Blay-Palmer, 2018).

At the global scale, there are several assessment tools that are either directly or indirectly linked to all or some dimensions of sustainable food systems. Arguably the most high profile in recent years have been the United Nation's Millennium Development Goals (MDGs) and the SDGs. MDG Goal 1 referred directly to food by calling for the eradication of hunger and poverty while the other seven goals were indirectly linked to improved food system sustainability through education, health, gender equality, environmental health, and building partnerships.

The SDGs build from the MDGs and are founded on 17 goals with associated targets that have been elaborated into 167 targets. SDG 2 (Zero Hunger) and the related targets provide countries with the opportunity to report on various dimensions of sustainable food systems, including food security and nutrition, productivity and incomes of small-holder farmers and other small-scale food-getters, land access, wild harvesting, sustainable production, protection of genetic diversity, and the correction and elimination of trade distortions. SDG 2 links to the other 16 SDGs with particular articulation with eradicating poverty (SDG 1), good health and well-being for all (SDG3), gender equality (SDG5), clean water and sanitation (SDG6), decent work and economic growth (SDG8), responsible production and consumption (SDG12), and climate action (SDG13).

The SDG targets are considered to be a blessing and/or a curse (Death & Gabay, 2015). In theory, the SDGs are an improvement over the MDGs in that they include all countries (not just countries in the Global South). Among concerns about the SDGs is the privileging of scientific knowledge, the disregard for traditional knowledge, the reinforcement and expansion of neo-liberalism, and the technocratization of sustainability assessment (Prato, 2016; Fukuda-Parr & McNeill, 2019).

Sustainable development practice has also prompted reflections about new or improved forms and processes of information gathering and information diffusion. Conventional scientific expertise, which is the most evident source of information when managing public action, shows some limits when applied to the challenge of providing information for mobilizing decisions (Death & Gabay, 2015). Other examples of global-scale data collection related to food systems include comparative assessments that are undertaken for narrower foci, for example the Prevalence of Undernourishment (PoU) as part of the annual State of Food Insecurity (SOFI). (For a critique of changes to the SOFI process in 2012 refer to *Framing Hunger*, Lappé et al., 2013).

Sustainable food assessments can also be understood through a thematic lens, for example using policy, food sovereignty, and/or food security as the basis for an assessment. On the policy side, Perez-Escamilla et al. (2017) evaluate the usefulness of national and global food security indicators as they make the case for SMART (specific, measurable, achievable/attainable, relevant and time-bound) indicators that facilitate evidence-based policy-making. Schader et al. (2016) also apply SMART criteria to link farm-level activity with broader sustainability goals. Using SAFA, they assess synergies and trade-offs between sustainability goals and find, in particular, that environmental dimensions are often traded-off for economic benefits. They make the case that policy needs to take into account these trade-offs. Both of these papers build on previous work such as that by Bauler (2012) who developed a framework to analyse the usability of environmental sustainability indicators for policy-makers, understanding that defining or implementing metrics is effectively a political act. Bauler (2012) argues for the 'initiation of a "politics of policy indicators" and focuses on an expanded usability analysis of indicators that includes both an LCS (legitimacy-credibility-saliency) framework and a conceptualisation of indicators as "boundary institutions"' (p. 43) that work at the interstices of three possible pairs of policy actor arenas: science–policy, policy–society, society–science. Bauler's (2012) ultimate concern is how indicators can be developed to have the highest usefulness profile so that quality information is gathered and diffused in ways that foster robust policy decisions towards sustainability. In keeping with IAASTD and other initiatives, he questions the ability of traditional forms of scientific/modernist knowledge to be usable across policy actor groups or to represent important and changing data in ways that are effective for policy decision-making. Instead, he suggests that many existing indicators, as a basis for policy creation, are problematic given challenges previously mentioned including privileging modernist science as the primary expertise, as well as questions of accuracy and how information is made available (Death & Gabay, 2015).

In other cases, the goal is to create alternative framings and guidelines in support of sustainable food system transformation. Binimelis et al. (2014) use food sovereignty as a normative framework to elevate the profile and possibilities for SFS (see also Levkoe & Blay-Palmer, 2018) through rural development and agroecology. Their environmental scan resulted in five categories (access to resources; production models; transformation and trade;

food consumption and the Right to Food; and agricultural policy and civil society organizations) that can be applied at an international scale using 128 indicators. They then adopt a local focus through their review of the Catalan food sovereignty assessment (based on Badal et al., 2011). Using three methods – documentary analysis, interviews, and participatory action research – data were gathered that allowed a 'self-reflexive analysis' (Badal et al., 2011, p. 333) of food sovereignty movement goals and their alignment with the 'actually-existing' (Brenner, 1997) food system. Along with other assessment tools and related processes such as the Sustainable Cities initiative in the UK (Moragues et al., 2013), this project is an excellent example of assessments being used for community and consensus building, both locally and at higher scales. The application of food sovereignty goals in Catalan exposes the place-specific nature of these projects on the ground, as assumptions inherent to La Via Campesina and food sovereignty, developed primarily in the Global South, required reinterpretation for Catalan. The application of a modified food sovereignty approach is also the case for the *Food Counts* report card in Canada that includes 'Food is Sacred' as a category, drawing on work by the national organization Food Secure Canada and its People's Food Policy.

Clearly, then, sustainable food system assessment is important for understanding issues at a given scale and to enable comparisons between cities, regions, and countries as well as to enable knowledge sharing and learning. Assessment can also clarify the role and directions for policy and raises important questions about how the information is shared. It is equally essential to be aware of the power that is activated through assessment with a view to enabling sustainability. Assessments can be used to foster agency for civil society, shine a light on inequities and environmental degradation and where there is room for improvement in support of sustainability, or not.

With this overview in hand, we now turn to a brief description of the chapters in this book, proceeding from conceptual considerations to how assessments can be operationalized and ending with some examples of impacts and outcomes of sustainable food systems research.

Introducing the book chapters

Recalling the insights from the collective research referred to earlier in the chapter, the three overarching themes used to organize this book are: conceptual considerations, operationalizing metrics, and project goals and outcomes.

Part I of the book includes three chapters that focus on conceptual foundations including sense-making and vision, and place. Chapter 2 by Valette et al. examines emerging research for a participatory mapping methodology of innovation impact pathways as they occur in sustainable urban food systems. The URBAL (Urban-Driven Innovation for Sustainable Food Systems) project focuses on innovations in consumer practices, value-chains, and governance by examining 12 case studies – referred to as Urban Food Innovation Labs (UFILs) – in eight cities in the Global South and North. The research begins from the visions of innovators in each UFIL and recognizes that the methodology needs to accommodate unique place-based dimensions. As a result, URBAL raises questions about how to develop a methodology that is at once generic enough to be used to study all urban food system innovation in the context of sustainability dimensions, and at the same time allow for useful insights in each place where the approach is applied.

The next chapter by Spring et al. takes us to northern Canada and describes the food systems of communities throughout Canada's northern boreal forest and their reliance on the health of the surrounding land, its animals, and waters. Traditional foods – those gathered, fished, and harvested from the land – remain the basis of these community food systems. However, climate change is having profound negative impacts on both availability and accessibility to traditional food sources through its impacts on the ecological integrity of the boreal system. Combined with other socio-economic factors, this has increased dependency on processed food purchased through the stores and has contributed to the rise of food insecurity across the region. This chapter discusses the lack of assessment tools for the complex food systems found in northern communities, and the need for more holistic approaches to measurement that connect food systems, land, and ecosystem health. The authors discuss research being conducted with, and in, the community of Kakisa, Northwest Territories, that highlights a community-driven assessment to understand the health of the boreal forest ecosystem. Kakisa, like many northern communities, see the health of the boreal forest ecosystem as the most important part of their food system. The well-being of northern communities therefore relies on the resiliency of the boreal forest and the ability of all aspects of the food system to support the overall form and function of the ecosystem.

In Chapter 4, the final chapter in Part I, Meter considers how a complex-adaptive-systems approach has simplified assessments performed in 140 US regions. The intent is to foster more self-reflective and effective action by framing the food system to be assessed as a complex adaptive system, which is a system that the component entities are learning from each other and adapting to what they learn. As Meter explains, introducing a conscious approach to adaptive complexity reaps several rewards. First, it harnesses quantitative and qualitative data to illuminate core system dynamics. Second, it taps essential wisdom from community members, helping to identify the essential levers that can move a given food system to greater sustainability. Third, by so doing, this approach facilitates community foods initiatives that develop *systemic* theories of change, for which indicators cut across issues that are often separated by disciplinary divides. These suggest measurement approaches that illuminate how system dynamics have shifted, and whether systems levers have moved, rather than focusing solely on programmatic outcomes.

Meter's chapter provides an excellent segue to Part II on operationalizing sustainable food system assessment where we explore the complexity of metrics specifically when developing frameworks and representations, the challenges of working within and across scales, and dealing with practical constraints such as data availability. In the first chapter of Part II, Battersby describes the process of writing the 2013 food system and food security study commissioned by the City of Cape Town, South Africa. This study honed in on data gaps and what they reflect about the politics of food, highlighting both scale-related governance and data challenges. Given the historical lack of consideration of food as a local government competence, researchers found that available data were either poorly disaggregated or unavailable, requiring the state to depend on poorly matched proxy indicators. This chapter reflects on the process, focusing on the politics and practicalities of data collection and disaggregation for municipal and regional food system assessments. Building on this experience, the author then considers the current challenges faced by local governments reporting against externally generated indicators (such as the UN SDGs and the 100 Resilient Cities indicators), which are time-consuming to report against at best, and are often at odds with the logic of local development plans. The chapter concludes with an argument for more micro-scale, collaborative, community-generated data and assessment frameworks.

Moragues-Faus's chapter addresses questions raised by the research in Cape Town, recognizing the need for more structural change through policy to enable SFS-based transformation. Complementing concerns expressed by Battersby, the author describes the explosion of indicator-based research and the associated ubiquitous tools to measure and facilitate sustainable development in food systems. Moragues-Faus critically questions the usefulness of these initiatives, including academic and practitioner-led processes, bottom-up and top-down approaches, and a local-, national- and international-level focus. Questions such as who can use these tools and for what purpose become central since these measurements ultimately produce specific narratives of how foodscapes function and develop. To begin to answer these questions, this chapter provides an analysis of an Action Research (AR) process to measure progress in UK cities towards more sustainable food futures. After presenting the action research process and associated methodology, the author assesses the different co-production and reflective practices at play as the indicators are developed. The chapter assesses the benefits and limitations of the AR approach to support transitions towards more sustainable and fair urban food systems. The author concludes by encouraging both practitioners and researchers to embed reflexive and co-productive governance principles within their everyday practice.

The final chapter in Part II, Chapter 7 by Prosperi, Allen, and Cogill, addresses operationalization questions that tested an iterative methodology to develop consensus about a set of food systems indicators at the nexus of broad sustainability goals and local contexts. The project sought to develop better knowledge of food system dynamics to understand more about how to strengthen both food and nutrition security outcomes and improve the social, environmental, and economic performance of food systems. The goal was to support policymakers, researchers, and local practitioners who strive to identify and develop SFS indicators. The project was led by Bioversity International and the International Center for Advanced Mediterranean Agronomic Studies-Mediterranean Agronomic Institute of Montpellier (CIHEAM-IAMM) to build consensus among the multidisciplinary food research community. The set of indicators (specific to the Mediterranean context) was developed to assess the different SFS dimensions with the view to inform policymakers. Applying a Delphi survey, consensus was reached after three rounds of consultations involving 52 experts for 14 indicators. The application of the Delphi method in the study provides several lessons to enhance validity, replicability, participation, and consensus for further surveys. The chapters in Part III make it clear that implementing useful place-based sustainable food system assessment tools is challenging and that engagement is key to developing and using tools and processes that are relevant.

Part III of the book builds from previous insights into the importance of engagement by focusing on the impacts and outcomes of SFS assessment and metrics, including policy generation, community engagement, embedding change in communities, and knowledge dissemination. There is clear overlap here with Part II of the book, since the toolkits mentioned below are simultaneously examples of operationalizing frameworks. This part begins with Chapter 8, a contribution by Palmer and Santo that describes a toolkit developed to measure advocacy capacity as a foundation to grow food policy. The underlying insight for the toolkit recognizes that while food policy councils (FPCs) aim to foster more just, healthy, and sustainable food systems by discussing, shaping, and assessing food system policies and programmes in their communities, many struggle to understand how to engage effectively in advocacy work. Recognizing this common challenge, the authors used an iterative consultation to create a comprehensive online assessment toolkit that helps build the capacity of stakeholder groups like FPCs to influence local and state level food policy. The toolkit will be most useful for groups working along the continuum of advocacy – those that aspire to do more advocacy in the food system arena, those that are struggling with their advocacy work, and those that are heavily involved in advocacy and want to evaluate and improve their efforts. The toolkit includes a self-assessment that provides a sequence of activities to help FPCs better understand the advocacy and policy process, evaluate their current advocacy capacity, and/or use the results to guide discussions about how to get started. This chapter explores the goals, development, and content of the toolkit, with specific attention to equity and systems-thinking metrics. It reflects on how the toolkit has been used in practice by one FPC in Lynchburg, Virginia, USA.

Also focused on the sub-national scale, Chapter 9 by Santini, Dubbeling, and Blay-Palmer describes the City Region Food System (CRFS) approach to food system assessment developed by the UN-FAO and RUAF, with support from the LCSFS, in seven pilot city regions. The chapter describes the CRFS toolkit that was derived from reviews of the process of change and lessons learned in the pilot projects. Elements of the toolkit include sections on how to define the boundaries of a CRFS, how to create a local vision, how to characterize and conduct an analysis of a CRFS, and how to engage in policy and planning. The toolkit also suggests methodologies for engaging in governance and multi-stakeholder dialogue processes. The toolkit is linked to synthesis reports for each of the seven pilot cities, providing details about both the implementation and results of these CRFS interventions. The CRFS approach enables concrete policy and investment opportunities within which complex sustainable development issues can be addressed, and through which rural and urban areas and communities can be directly linked. In turn, improved CRFS may help achieve better economic, social, and environmental conditions in both urban and surrounding rural areas. As with the toolkit developed by Palmer and Santo, the CRFS toolkit provides both the opportunity for self-assessment and points to ways forward for city regions.

The next chapter, Chapter 10, co-written by academics and practitioners and led by Paredes and Cole, takes us to the micro scale through the lens of households and individuals in Ecuador. The research explores the premise that to move towards food system sustainability individuals must play a larger role by becoming more responsible food consumers. Based on diverse literatures and observed practices, the multidisciplinary Ekomer research team conceptualized responsible food consumption, designed a Responsible Consumption Index (RCI), and applied it in a sample of randomly selected households in three city regions, or counties of Ecuador (Ibarra, Quito, and Riobamba), and in samples of individuals shopping at agroecological food outlets or fairs (open air markets). They describe the heterogeneous distribution of the RCI dimensions and overall score across different populations as well as the analysis from their market surveys. Their preliminary analyses show a positive relationship of RCI with daily fruit and vegetable consumption, as well as table-salt reduction practices. From the open-air market research they identify consumption of Andean grains, direct purchasing from producers especially of agroecologically produced food as more in line with SFS. The authors conclude by reflecting on using these insights to strengthen sustainable and healthier food regimes in city regions. This chapter highlights the potential for enhanced agency of households and individuals and how they can help to enable change.

The final chapter by Nevin Cohen, Chapter 11, considers how to integrate both upstream and downstream food metrics. The research begins from the insight that food metrics typically measure the outcomes of food policies and programmes, not the root causes that create food system injustice, like poverty and discrimination. As a result, attention is focused on downstream interventions and the potential to address upstream determinants is often masked. Cohen makes a compelling case that integrating the two can enable planners to design better food policies and foster stronger collaboration among food and social justice advocates. The chapter illustrates the value of integrating upstream determinants and downstream food metrics with three examples from New York City: the role of federal immigration policies on access to federal food benefits; the role of zoning changes on real estate development that leads to food gentrification; and the effects of policies to increase wages and improve working conditions on low-wage food sector workers. The chapter concludes with strategies for iteratively integrating upstream and downstream metrics. These involve aggregating existing public data and emerging sources of big data in food planning, and encouraging urban planners to incorporate upstream data in their food planning activities.

This edited volume concludes with Chapter 12, which provides a reflection on the insights from the book and implications for future research and assessment for both theory and practice. No single approach is endorsed in this volume; rather, a diversity of approaches that can be adapted to diverse issues and contexts is highlighted. Together, these chapters provide valuable insights into the place-based challenges of navigating the multiple and often conflicting needs of practitioners, policymakers, and funders across multiple scales and the pressures that emerge as these tensions are addressed. By examining various assessment tools, the authors question both the relevance and politics of sustainable food systems by engaging directly with the motivations and processes that have emerged in response to the call for assessment approaches. Collectively, the authors are working to make indicators and tools that reflect, without diminishing, and that are relevant for the complex work of transformation towards increasingly sustainable food systems.

Acknowledgements

We would like to acknowledge the support of the Social Sciences and Research Council of Canada Partnership Grant, and the Laurier Centre for Sustainable Food Systems. We are also grateful to our community partners for their ongoing engagement and support.

Note

1 Chapter cross-references guide readers to similar concepts in the book that can be located using a key word search, given this book is open access and available online.

References

Anderson, M.D. (2015) The role of knowledge in building food security resilience across food system domains. *Journal of Environmental Studies and Sciences*. 5(4), pp. 543–559.

Badal, M., Binimelis, R., Gamboa, G., Heras, M., & Tendero, G. (2011) *Arran de terra. Indicadors participatius de Sobirania Alimentària a Catalunya*. Barcelona, Associació Entrepobles i Institut d'Economia Ecològica i Ecologia Política.

Battersby, J. (2011) Urban food insecurity in Cape Town, South Africa: An alternative approach to food access. *Development Southern Africa*. 28(4), pp. 545–561.

Bauler, T. (2012) An analytical framework to discuss the usability of (environmental) indicators for policy. *Ecological Indicators*. 17, pp. 38–45.

Binimelis, R., Rivera-Ferre, M.G., Tendero, G., Badal, M., Heras, M., Gamboa, G., & Ortega, M. (2014). Adapting established instruments to build useful food sovereignty indicators. *Development Studies Research*. 1(1), pp. 324–339

Blay-Palmer, A., Sonnino, R., & Custot, J. (2015) A food politics of the possible? Growing sustainable food systems through networks of knowledge. *Agriculture and Human Values*. 33(1), pp. 27–43.

Blay-Palmer, A., Santini, G., Dubbeling, M., Renting, H., Taguchi, M., & Giordano, T. (2018) Validating the city region food system approach: Enacting inclusive, transformational city region food systems. *Sustainability*. 10(5), p. 1680.

Born, B. & Purcell, M. (2006) Avoiding the local trap: Scale and food systems in planning research. *Journal of Planning Education and Research*. 26(2), pp. 195–207.

Brenner, N. (1997) Global, fragmented, hierarchical: Henri Lefebvre's geographies of globalization. *Public Culture*. 10(1), pp. 135–167.

Bricas, N. (2017) Les enjeux de l'urbanisation pour la durabilité des systèmes alimentaires. In: Brand, C. et al. (Eds.) *Construire des politiques alimentaires urbaines. Concepts et démarches*. Paris, Editions Quae, pp. 19–38.

Brundtland, G.H., Khalid, M., Agnelli, S., & Al-Athel, S. (1987) *Our common future*. New York, Oxford University Press.

Calori, A., Dansero, E., Pettenati, G., & Toldo, A. (2017) Urban food planning in Italian cities: A comparative analysis of the cases of Milan and Turin. *Agroecology and Sustainable Food Systems*. 41(8), pp. 1026–1046.

Crush, J., Frayne, B., & Pendleton, W. (2012) The crisis of food insecurity in African cities. *Journal of Hunger & Environmental Nutrition*. 7(2–3), pp. 271–292.

Death, C. & Gabay, C. (2015) Doing biopolitics differently? Radical potential in the post-2015 MDG and SDG debates. *Globalizations*. 12(4), pp. 597–612.

Dubbeling, M., Santini, G., Renting, H., Taguchi, M., Lançon, L., Zuluaga, J., De Paoli, L., Rodriguez, A., & Andino, V. (2017) Assessing and planning sustainable city region food systems: Insights from two Latin American cities. *Sustainability*. 9, p. 1455. doi:10.3390/su9081455.

Fawcett, S.B., Boothroyd, R., Schultz, J.A., Francisco, V.T., Carson, V., & Bremby, R. (2003) Building capacity for participatory evaluation within community initiatives. *Journal of Prevention & Intervention in the Community*. 26(2), pp. 21–36. Available at: https://doi.org/10.1300/J005v26n02_03.

Feenstra, G.W. (1997) Local food systems and sustainable communities. *American Journal of Alternative Agriculture*. 12(1), pp. 28–36.

FIAN (2016) *FIAN International Annual Report 2016*. Available at: https://www.fian.org/fileadmin/media/Publications/Annual_Report/FIAN_International_Annual_Report_2016_english_march.pdf.

Freedgood, J., Pierce-Quiñonez, M., & Meter, K. (2016) Emerging assessment tools to inform food system planning. *Journal of Agriculture, Food Systems, and Community Development*. 2(1), pp. 83–104.

Fukuda-Parr, S. & McNeill, D. (2019) Knowledge and politics in setting and measuring the SDGs: Introduction to special issue. *Global Policy*. 10, pp. 5–15.

Grenz, J., Thalmann, C., Schoch, M., Stämpfli, A., & Studer, C. (2011) *RISE (Response-Inducing Sustainability Evaluation)*. Version 2.0. Available at: https://saiplatform.org/uploads/Modules/Library/What%20is%20RISE%202.pdf.

Hinrichs, C.C. (2014) Transitions to sustainability: A change in thinking about food systems change? *Agriculture and Human Values*. 31(1), pp. 143–155.

Hinrichs, C.C. (2016) Conceptualizing and creating sustainable food systems: How interdisciplinarity can help. In: Blay-Palmer, A. (Ed.) Imagining sustainable food systems. London, Routledge, pp. 17–36.

Hipel, K.W., Fang, L., & Heng, M. (2010) System of systems approach to policy development for global food security. *Journal of Systems Science and Systems Engineering*. 19(1), pp. 1–21.

IAASTD (2009) *Agriculture at a Crossroads: Global Report*. International Assessment of Agricultural Knowledge, Science and Technology for Development. Available at: https://archive.org/stream/fp_Agriculture_at_a_Crossroads_Global_Report_English/Agriculture%20at%20a%20Crossroads_Global%20Report%20%28English%29_djvu.txt.

Ingram, J. (2011) A food systems approach to researching food security and its interactions with global environmental change. *Food Security*. 3(4), pp. 417–431.

IPES-Food (2015) *The new science of sustainable food systems: Overcoming barriers to food systems reform*. International Panel of Experts on Sustainable Food Systems.

Knezevic, I., Blay-Palmer, A., Levkoe, C.Z., Mount, P., & Nelson, E. (Eds.) (2017) *Nourishing communities: From fractured food systems to transformative pathways*. New York, Springer.

Lappé, F.M., Clapp, J., Anderson, M., Broad, R., Messer, E., Pogge, T., & Wise, T. (2013) How we count hunger matters. *Ethics & International Affairs*. 27(3), pp. 251–259.

Levkoe, C.Z. & Blay-Palmer, A. (2018) Food counts: Food systems report cards, food sovereignty and the politics of indicators. *Canadian Food Studies/La Revue canadienne des études sur l'alimentation*. 5(3), pp. 49–75.

McKeon, N. (2015) Global food governance in an era of crisis: Lessons from the United Nations Committee on World Food Security. *Canadian Food Studies/La Revue canadienne des études sur l'alimentation*. 2(2), pp. 328–334.

Meadows, D.H. (2008) *Thinking in systems: A primer*. Chelsea, VT, Chelsea Green Publishing.

Meul, M., Van Passel, S., Nevens, F., Dessein, J., & Rogge, E. (2008) MOTIFS: A monitoring tool for integrated farm sustainability. *Agronomy for Sustainable Development, Springer Verlag/EDP Sciences/INRA*. 28(2), pp. 321–332.

Moragues, A., Morgan, K., Moschitz, H., Neimane, I., Nilsson, H., Pinto, M., Rohtacher, H., Ruiz, R., Tisenkops, T., & Halliday, J. (2013) *Urban food strategies: The rough guide to sustainable food systems*. Document developed in the framework of the FP7 project FOODLINKS (GA No. 265287).

Neubert, S. (2010) *Description and examples of MAPP method for impact assessment of programmes and projects*. German Development Institute. Available at: http://www.ngo-ideas.net/mediaCache/MAPP/MAPP-Description.pdf.

Perez-Escamilla, R., Gubert, M.B., Rogers, B., & Hromi-Fiedler, A. (2017) Food security measurement and governance: Assessment of the usefulness of diverse food insecurity indicators for policy makers. *Global Food Security*. 14, pp. 96–104.

Prato, S. (2016) SDG 2: Facilitating corporate capture or investing in small-scale sustainable agriculture and agroecology? In: *Spotlight on Sustainable Development 2017*. pp. 40–48. Available at: https://www.2030spotlight.org/sites/default/files/download/Spotlight2017_2_2_Prato.pdf.

Schader, C., Baumgart, L., Landert, J., Muller, A., Ssebunya, B., Blockeel, J., Weisshaidinger, R., Petrasek, R., Mészáros, D., Padel, S., & Gerrard, C. (2016) Using the sustainability monitoring and assessment routine (smart) for the systematic analysis of trade-offs and synergies between sustainability dimensions and themes at farm level. *Sustainability*. 8(3), p. 274.

Stroink, M.L. & Nelson, C.H. (2013) Complexity and food hubs: Five case studies from northern Ontario. *Local Environment*. 18(5), pp. 620–635.

Part I

Conceptual foundations

2 An emerging user-led participatory methodology

Mapping impact pathways of urban food system sustainability innovations

Elodie Valette, Kerstin Schreiber, Damien Conaré, Veronica Bonomelli, Alison Blay-Palmer, Nicolas Bricas, Denis Sautier, and Olivier Lepiller

Introduction

Given the growing pressures on our planet by the current food system, it is increasingly important to understand the transformative potential of urban food systems and their capacity to build pathways to sustainability. According to the United Nations, more than half of the world's population now lives in urban areas with predictions that this proportion will increase to approximately two-thirds by 2050. Given current lifestyle and consumption practices, people living in urban areas monopolize three-quarters of all natural resources and account for 60–80 per cent of global GHG emissions (UNEP, 2011; Chapter 1, this volume).

But if cities concentrate sustainability problems, they are also (like rural communities) places of innovation that can contribute to building more sustainable food systems. Both urban government and citizen food initiatives flourish in Northern and Southern cities and offer new ways of feeding cities and connecting actors of urban and rural territories around food issues (Rocha & Lessa, 2009; Viljoen & Wiskerke, 2012; Robineau, 2015; Blay-Palmer et al., 2016; Brand et al., 2017; Blay-Palmer et al., 2018). Largely supported by civil society or local governments, these initiatives are usually presented as alternatives to the dominant food system (Lang, 1999; Moragues-Faus & Marsden, 2017) and ways to contest agro-industrial capitalism.

However, few research activities, so far, have aimed to analyse the effect of urban-driven changes occurring across both transitioning and industrialized countries on diverse sustainability dimensions. While numerous innovative initiatives are generated by cities – either by the public sector (Friedmann, 2007; Reynolds, 2009; Mah & Thang, 2013; Laidlaw, 2015), private sector (Brand, 2015; Reardon, 2015), or by consumers themselves (Wertheim-Heck et al., 2014), their actual effects on sustainability have been insufficiently documented (see Chapter 9, this volume). Moreover, while there is an implicit positive causal relationship between urban innovations and sustainability, most of these relationships remain undocumented and unclear.

Among them, for example, is the idea that local food trade is inherently environmentally friendly (Katz, 2010) or socially just. Yet many works show the complexity and relative truth of this assertion (Born & Purcell, 2006; Desrochers & Shimizu, 2008). Some of these innovations may also have negative impacts on social issues. Slocum et al. (2016) pointed out how, in spite of the will to create and inhabit more equitable food spaces, the food movement often fails to lead to food justice, instead reinforcing existing race, class, and gender inequalities (Clancy, 1994; Freidberg, 2003). In a nutshell, urban food innovations do not necessarily address sustainability issues.

Additionally, existing initiatives that focus on indicators, such as the UN Sustainable Development Goals, privilege technocratic approaches and data, while case-study analyses can be limiting in terms of their transferability (see Chapters 1 & 12, this volume). Within this context, understanding how to move towards increasingly sustainable city region food systems is a challenge. The research programme URBAL (Urban-driven Innovations for Sustainable Food Systems) (2018–2020), funded by Agropolis Fondation (France), Fondation Daniel & Nina Carasso (France/Spain), and Fundazione Cariplo (Italy), and coordinated by CIRAD (France) and the Laurier Center for Sustainable Food Systems at Wilfrid Laurier University (Canada), seeks to build and test a participatory methodology to identify and map the impact pathways of urban-driven innovations on all the dimensions of food systems sustainability. By testing this methodology through various case studies internationally – Urban Food Innovation Labs (UFILs), including sites in the Global South and North – this project aims to provide decision makers with information on how innovations can contribute to, or work against, building more sustainable food systems, thus assisting them to determine which actions should or should not be taken (see Chapters 3 & 9, this volume).

This chapter presents the general framework for the URBAL project as well as the main interwoven considerations and approaches that are the backbone of the methodology. Please note that this is an ongoing project and that it has evolved since the chapter has been written. We will point out some changes in the methodology as the chapter proceeds.

How to assess the impact of urban-driven innovations on the sustainability of food systems?

The main objective of the URBAL project is to provide urban policymakers, urban innovators, and funders with a low-cost, easy-to-implement, and context-adaptable methodology that can be used as a robust tool to make the impact pathways of innovations on all the dimensions of the sustainability of food systems more explicit. Its purpose is not to provide evaluation per se, but rather to help uncover the ways social innovations link with sustainability. Hence URBAL builds from existing work on different topics: impact pathway mapping, innovation with an emphasis on social innovation, participatory action research, and sustainable food system assessment.

In the field of food system assessment, numerous implementation projects and research efforts have aimed at identifying relevant indicators to assess food system sustainability. These range in number from a few to more than one hundred indicators that strive to address all the dimensions of sustainability (Singh et al., 2012). For example, FAO and RUAF Foundation (Holland), associated with the Center for Sustainable Food Systems at Wilfrid Laurier University (Canada), developed a set of indicators to map and assess city region food system sustainability (see Chapter 9, this volume). This tool has been tested on several city regions in the world. The Milan Urban Food Policy Pact has also developed a set of indicators to measure urban food system sustainability.

These are nevertheless time- and money-intensive methodologies that cannot be used easily by local authorities or innovators to better inform their decisions and actions in a context where budgets are limited. This leads many policymakers to plan without a clear idea of the impact they can expect (Jobert & Muller, 1987; Callon et al., 2001). For these reasons, it is necessary to explore simpler and participatory methodologies that can foster social learning in the context of public policy (Bennett & Howlett, 1992; Hall, 1993; Rose, 1993).

Mainly quantitative evaluation methodologies

Reliable analysis and evaluation techniques (Patton, 2012), including 'utilization-focused evaluation', are applied widely. Yet, work on the link between food systems and holistic sustainability is scarce (Aubin et al., 2013) and almost non-existent in relation to social innovations and impact pathways. That said, there has been some progress. Popov et al. (2017) have analysed social innovation using a quantitative approach to measure impacts of social innovation. Dhondt et al. (2016) developed a socio-economic impact analysis. Wiek et al. (2017) use log modelling to analyse sustainability experiments in cities, forming the foundation for the evaluation scheme of Luederitz et al. (2017), who integrate sustainability transitions. Luederitz et al. looked specifically at urban sustainability transition labs (USTL) and utilized log models to assess USTL outputs and outcomes, both site specific and city-wide. Studies have also looked into whole food systems (Institute of Medicine and National Research Council, 2015), but local initiatives in relation to food were not taken into account. The development of assessment tools for measuring and mapping social food innovations and their contribution to sustainability and systemic change still remains incomplete (Aubin et al., 2013; see also Chapters 4 & 7, this volume).

Multiple evaluation methodologies rely on standardized, expert-led, and quantitative data collection and analysis tools to ensure comparability, reproducibility, and objectivity. For instance, such methods are chosen in fields like agriculture, health, and water management (Peterson, 2006) under the assumption that these provide sufficient data and advice for unbiased and scientifically informed decision-making (Leach et al., 2010). On the other hand, Naudet et al. (2012) argue that quantitative and financial evaluations neglect crucial elements, underlining their inability to establish the cause and effect relationship that allows one to assess development policies or any kind of project.

Aubin et al. (2013) conducted a critical review of sustainability evaluation methods and their applicability in the food system, consisting of common methods ranging from Life Cycle Assessment (LCA); economic methods such as cost-benefit analysis (CBA); retrospective assessment of food uses and resources; physical and synthetic indicators of environmental pressures; nutritional and epidemiological approaches; methods to assess obstacles to, and motivations for, sustainable consumption; composite social and ethical indicators; and multi-criteria and participatory assessment tools. Their main critiques are that prevailing methods are too global (input–output analysis); too local (LCA); too single-dimensional (physical and synthetic indicators); too static (LCA, economic methods, physical, and synthetic indicators); too predictable and narrow in methodology (LCA); too imprecise (LCA); require data that is difficult to acquire (input–output); are vulnerable to errors (input–output); or too focused on macro-nutrients (food availability) and consumer motivation (assessment of sustainable consumption); or do not take informal economic trade into consideration (input–output). This latter concern is especially relevant in developing and emerging countries. At the same time, Aubin et al. (2013) deny qualitative or context-specific elements or, for instance in the case of cost–benefit calculations, undervalue potential future dynamics (Howarth & Norgaard, 2013) in favour of the focus on current value flows (Portney & Weyant, 1999). Lastly, focusing on effect measurement methods, they neglect to scrutinize causal relationships (Mayne, 2011).

The various dimensions of sustainability

Sustainable development, as introduced in the Brundtland report (World Commission on Environment and Development, 1987), has significantly influenced the demands on, and the design of, assessment tools that must deal with contextual (site-specific conditions) and temporal issues (impacts occur at various, often lengthy, times after the intervention), as well as multidisciplinarity (Temple et al., 2018). Based on these considerations, new and more qualitative approaches to sustainability assessment have been proposed to help urban food system governance. Some methods for identifying relevant sustainability indicators range from expert-led choices to more participatory processes that help communities identify their own indicators (Fraser et al., 2005; Abi-Nader et al., 2009; see also Chapters 1 & 12, this volume). For example, Landert et al. (2017) developed a participatory process to operationalize the Guidelines for Sustainability Assessment of Food and Agriculture Systems (SAFA) developed by the UN Food and Agriculture Organization (FAO). The multi-criteria assessment method uses 97 indicators evaluating 51 of the SAFA sub-themes. The choice of the indicators is based on a comprehensive approach of a sustainable food system and is based on policies and measures related to processes within the urban food system.

Very recently, Carlsson et al. (2017) have proposed a participatory methodology based on the Framework for Strategic Sustainable Development (FSSD) through a modified Delphi Inquiry Process to identify key indicators for tracking progress towards the success of the food system at a local level (see also Chapter 7, this volume). Both in Landert et al. (2017) and Carlsson et al. (2017), the methodology implementation has been made possible thanks to the availability of data. Two improvements on these approaches are worth noting. First, data availability might not be guaranteed in developing countries. Second, in both cases, methodologies aim to assess and improve the whole food system and its sustainability more than to drive and improve local innovations.

Finally, while several research teams have explored the assessment of sustainability innovation impacts on the food system, they often limit the assessment to one or two dimensions of sustainability: environmental issues (with LCA tools) and/or nutrition (through consumption surveys) (see Chapter 10, this volume). The interactions between sustainability dimensions have rarely been taken into account. In order to better address this gap, some assessment methodologies use participatory mapping to identify and explore impact pathways of innovations (Douthwaite et al., 2007a; Proietti et al., 2015). URBAL follows this path.

It's also about politics

Evaluation also has a political dimension. Since sustainability transition experiments are embedded within structures and power relations, self-critical reflexivity within an evaluation is required (Avelino & Rotmans, 2009). Adding the concept of complex adaptive systems (see Chapter 4, this volume), one must acknowledge that projects and programmes are embedded in political, institutional, social, and economic systems, which evaluators can use to understand how these systemic affiliations and structures hinder or foster causal chains.

Even though a more holistic perspective covering impacts on sustainability is desired (Joly et al., 2015), most approaches are nevertheless one-dimensional. More recent ones try to balance these methodological obstacles by analysing, for instance, networks, political and institutional arrangements, as well as the use and production of knowledge to help policymakers in the quest to negotiate between different, often conflicting or competing options and pathways to social improvement (Bozeman & Sarewitz, 2011; Blay-Palmer et al., 2018). Some studies have addressed these methodological drawbacks. For instance, a recent study by Sanyé-Mengual et al. (2018) aimed to approach the field of urban agriculture research assessment and investigated the linkages between sustainability dimensions in urban food production through a participatory evaluation.

Rather than picking any random methodology, it is crucial to stand back and reflect on the needs and interests of stakeholders, the questions they want to answer, and the available resources (Fawcett et al., 2003; Neubert, 2010; also see Chapter 1, this volume). Both internal and external factors must be considered, such as public relations for marketing and funding acquisition purposes (Berg et al., 2009), 'accountability requirements', the 'existence of procedural routines for social inquiry', as well as the 'nature of the sought knowledge', are pivotal in the decision process towards an epistemologically fitting assessment toolset (Dhondt et al., 2016, p. 22) or internal purposes like project management and learning (Berg et al., 2009). Therefore, the efficacy of the demanded methodology depends on customizability, in order to account for the diversity of numerous forces in play (Dhondt et al., 2016). Specifically for use in participatory settings, the methodology should be inclusionary and facilitate capacity building and learning, as well as be comprehensible and easy to apply (Springer-Heinze et al., 2003).

The URBAL methodology: change-based and participatory theory

Based on these theoretical and methodological contexts, the participatory methodology that is being developed uses two main interwoven considerations and approaches.

Impact pathways

The aim of the URBAL methodology is to help disentangle the goals and strategies that lead an innovation to sustainability, not necessarily to measure its impact. That is why we chose to work with impact pathway mapping. While the identification of impacts can give clues about the effects and implications of a programme, project, or other initiative, it cannot answer the question of how and why an impact has occurred. Evaluation must be capable of disentangling the goals of a programme and the strategies in place to reach them (McLaughlin & Jordan, 2015). Pawson et al. (2005) emphasize the importance of outcome-generating mechanisms, causality, and the comprehension of both in their 'generative model of causality': 'to infer a causal outcome (O) between two events (X and Y), one needs to understand the underlying mechanisms (M) that connect them and the context (C) in which the relationship occurs' (Pawson et al., 2005, p. 22).

In the process of planning, organizations that perform social innovations often define specific goals they want to reach in response to the social construction of a problem. These goals are ideally tied to immediate targets, yet are long-term-oriented, clearly formulated, and measurable. More generally, innovators are able to draw the proposed path towards achievement of their stated goals. Doing so, organizations articulate external and internal economic, political, environmental, and social drivers that can alter the trajectory towards their chosen goals. They construct (implicitly or explicitly) several pathways and choose the one with the highest estimated efficiency (OECD, 2013). This logic can be defined as a theory of change, which proposes an explanation about how the actions and activities of a programme, intervention, or project are intended to contribute to planned effects and the reasons and mechanisms that link the operation (Mayne, 2011).

Impact Pathways Analysis (IPA) is a popular approach to investigate theories of change in the planning and evaluation process, using logic models to visualize elements and links for an intervention. IPA can provide a useful tool as it is dedicated to the questions of why and how a given intervention has led to an impact on whom, and which specific conditions were in play (GIZ, 2012; Quiédeville et al., 2017), by mapping and discussing these components (Dhondt et al., 2016).

IPA serves a descriptive function, that is, grasping the activities, circulation, transformation, and utilization of knowledge (Quiédeville et al., 2015) and other elements. Furthermore, it aims to provide an understanding of cause and effect relationships that have led to change (Reade, 2008) on the micro, meso, and macro levels (GIZ, 2012). The 'pathways approach' recognizes that these complex interactions can sometimes be self-reinforcing, strengthening dominant narratives and trajectories favoured by powerful actors, and crowding out alternatives favoured by marginal groups. Impact pathways analysis helps to challenge and scrutinize change processes and their dynamics (Reade, 2008). Particularly challenging can be documenting the evidence for undesired impacts, long-term impacts, and diffuse impacts, which do not address a specifically demarcated target group. While IPA has been applied in diverse contexts, it has mainly been carried out for research and development accountability programme evaluation to communicate with external funders and stakeholders. Other methodologies,[1] including ours, are more practitioner-oriented, where impact pathways are rather a management and strategy tool, that is, for internal use.

Participatory-based

Evaluative studies have often been conducted by researchers and other scientific experts. However, the approach to research *on* rather than *with* stakeholder groups, can lead to insufficient and biased conclusions that do not mirror the lived experiences of the impacted stakeholders. Many researchers therefore suggest using participatory approaches in order to address these shortcomings (Hummelbrunner, 2007; Meter, 2007; Berg et al., 2009). Multi-stakeholder planning has been described as especially crucial in complex agro-food systems (De Zeeuw & Dubbeling, 2015); the same can be applied to the evaluation perspective.

Participatory approaches are commonly chosen in impact pathways analysis since they allow a balance between the researchers' needs for specific information and the ability of participants to contribute their original experience. Cousins and Earl (1992) define participatory evaluation as 'applied social research that involves a partnership between trained evaluation personnel *and* practice-based decision makers, organization members with programme responsibility; or people with a vital interest in the program' (p. 399). In this case, the assessment takes place in a collaboration between stakeholders and experts, which also enables cross-comparison between statements (Berg et al., 2009). Hummelbrunner (2007) argues that this constellation in itself forms a system, the 'evaluation system', in which 'clients' and 'evaluators' influence each other's world views by intervening and reacting.

Participatory evaluation can have different operational benefits: it increases the usability and relevance of results into the everyday practices of participants (Cousins & Earl, 1992) through the appropriation of the evaluation process (Quiédeville et al., 2017), and the opportunity to raise locally relevant questions (Zukoski & Bosserman, 2017). Scholars also value participatory evaluation as very effective because of the possibility to adjust ongoing projects and programmes during execution, as well as to immediately recover from unexpected disruptions (Rodríguez-Campos, 2017). Participatory evaluation is not just the programme, project, or product, but also a process that enables dialectic discussions, which should lead to organizational learning (Suárez-Herrera et al., 2009), allow participants to express their opinions, have their voice and perspectives incorporated into the initiative (Marra, 2015), and provide systematic information to support learning and decision-making (Douthwaite et al., 2007b). This has been shown to support the integration of new ideas and constructs into existing mental maps and cognitive structures. The application of participatory evaluation and the stimulation of learning knowledge creation can also foster change (Plottu & Plottu, 2009; Marra, 2015). Scholars argue that capacity building is yet another important potential of participatory evaluation (Ekirapa-Kiracho et al., 2017).

A crucial part of the participatory evaluation methodology is the involvement of a wide and diverse range of key stakeholders (McLaughlin & Jordan, 2015; Zukoski & Bosserman, 2017), although the extent of involvement and the diversity of stakeholders can vary throughout the process at different stages (Rodríguez-Campos, 2017). Certainly, this diversity can render the evaluation difficult and make consensus impossible to achieve. Nevertheless, in the least case, it can generate a platform where conflicting perspectives can be laid open and discussed. Discussion and juxtaposition of the reasons and processes of change are the main purposes.

A three-stage process

Based on these statements, URBAL proposes to build and test a tool that can help different actors identify the potential and risks for different sustainability dimensions for urban food system innovations. Building from assumptions of participatory engagement, such as participatory budgeting (Cabannes, 2004) or participatory certification (Nelson et al., 2010), we will focus on policymakers and practitioners. The goal is to build a cognitive map or logical frame that makes explicit the impact of innovations on sustainability. These maps will identify the actual changes produced by the innovation on sustainability, the ways they are induced by the activities performed by the innovation, and the ways they interrelate, from short-term changes (outputs) to medium-term (outcomes) and to long-term changes (usually referred to as impacts). The chosen approach will therefore assess not only the intended and unintended impacts on all sustainability dimensions, but also the pathways that led to these changes. These pathways that help identify a theory of change for each innovation are not necessarily linear constructs: the interrelation, convergence, potential divergency between the various changes and pathways towards the different dimensions of sustainability, also build a systemic theory of change, emphasizing positive and negative feedback loops, unforeseen changes, and unforeseen contradictions between pathways, which we believe are particularly relevant to address the issue of the sustainable food systems.

The URBAL project aims to offer innovators, funders, and policymakers useful and clear information about the functionality and relevance of innovations. It aims at providing policymakers the tools to allocate funding and implement appropriate support mechanisms to foster sustainable change and prevent adverse societal effects. It also aims at providing innovators the tools to better assess their activities and to reframe their goals if they identify gaps and strengths. As such it will be of a participatory or collaborative nature (see also Chapter 8, this volume).

A key goal is to address all dimensions of food system sustainability. The sustainability framework developed by Bricas (2017) is at the core of URBAL's research approach so that we include economic, environmental, socio-cultural, food security, and nutrition and governance considerations as central to our research.

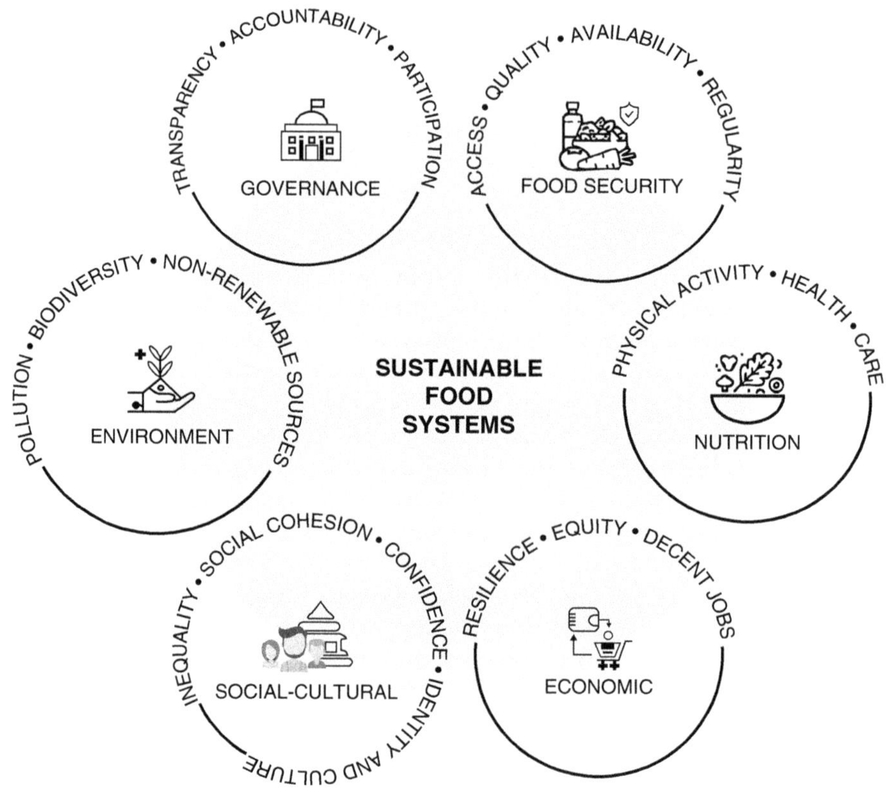

Figure 2.1 The dimensions of sustainable food systems.
Source: Adapted from Bricas, 2017.

As Figure 2.1 shows, the sustainability dimensions encompass five main areas:

- Social dimensions include social cohesion, inequality, confidence in the food system, as well as identity and culture;
- Economic dimensions include decent jobs, equity, and resilience considerations;
- Food security and nutrition dimensions are closely linked and include physical activity, health care as well as food access, availability, regularity, and quality;
- Environmental dimensions include pollution, biodiversity, and non-renewable resources;
- Governance dimensions include participation, transparency, and accountability.

A stakeholder-oriented methodology

As a general roadmap, we suggest a three-stage collaborative process (Figure 2.2), that will help document the innovation impacts. To consider long-established innovations as emerging ones (in which case it would be difficult to identify actual impacts due to the nascent nature of the innovation), and to dispel the fuzzy image that comes with the term 'impacts', one of the first methodological choices was to use the term 'change' and 'result' instead of 'impact'. In this context, outputs are the products or services generated by the innovation. Outputs can lead to direct observable changes and can be adopted and adapted by other actors in the same or other contexts. This is what we call 'Results' – short-term, medium-term or long-term changes linked to the innovation and its context, which are both influenced by and influence contextual factors.

The central question is 'How have ideas and/or practices changed because of the innovation?' In order to address this question, Step 1 will be dedicated to the collection of background information through interviews that will help to raise awareness about the innovation, document the context, and understand the motivation of the practitioners for the innovation. The main outputs of Step 1 will be (1) a chronogram, which displays visually the most relevant events in the genesis and the development of the innovation; (2) an actor network map that diagrams the system of actors and their mutual relationships; (3) a case description including the local and global context for the innovation; and finally (4) an impact pathway map that describes what strategies the project's stakeholders have used and what activities they have performed to bring about necessary actions to achieve the project vision, including unexpected short-term and long-term changes, as well as takes into account positive and negative feedback loops, as previously said.

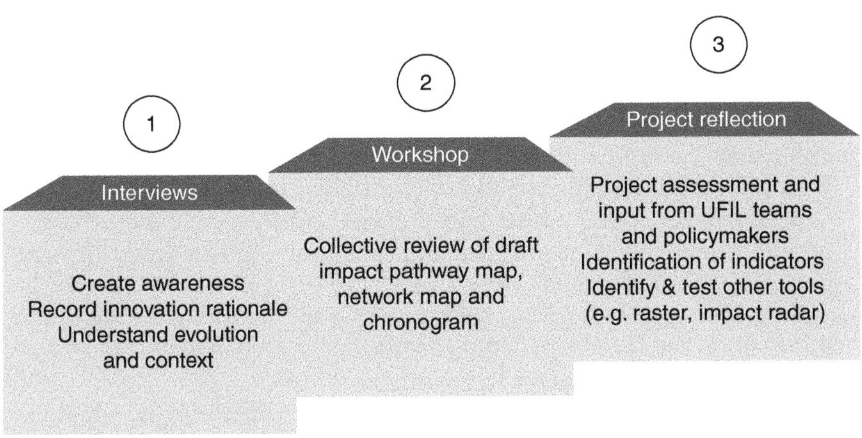

Figure 2.2 The three stages of the URBAL methodology.

This cognitive map is based on an Outcome Logic Model, which will be explained in the next section.

In Step 2, a workshop is organized involving stakeholders (practitioners, partners, users) to review the Impact Pathway map developed in Step 1 and discuss comments and improvements to the various pathways leading from the innovation to related changes. This step is dedicated to discussing and reflecting on the innovation's theory of change and evolution, which the innovators defined in Step 1. This step is crucial for the confirming or challenging the impact pathways identified in Step 1 through interviews and a literature review. It is also crucial for defining, discussing, and potentially eliminating causal linkages between each pathway step, and gaining awareness about the process and elements of the innovation (specifically policymakers, who can benefit from this knowledge in order to increase the efficiency of their decision-making). The selection of the workshop participants will depend on the sustainability dimensions and on the questions that will be addressed in depth in the workshop. As such, this selection strongly relies on the outputs of Step 1.

Step 3 is a meeting or a workshop to reflect on the results and the project as a whole with innovators, stakeholders, policymakers, and/or funders. This final step may also be used to identify indicators for benchmarking and actually measuring changes and impacts.

Stakeholders are the core actors in impact pathway mapping. It is advisable to involve 'wise practitioners', that is, people who are well immersed in the relevant area of interest but can also offer broader, more abstract contexts. Furthermore, food system innovations connect a multitude of different actors and stakeholders that should be categorized to provide a wide-ranging overview (Meter, 2007; see also Chapter 4, this volume) making the actor mapping in Step 1 so crucial.

The Impress method suggests the separation of actors into three categories (major actors, influential actors, and impacted actors) according to the role they play in the innovation process (Barret et al., 2018). This idea is used and adapted to URBAL, which proposes the following four categories:

1. Innovation holders: ideally founding members of the innovation.
2. Policymakers: representatives from municipal or regional governments with policymaking capacity, ideally with previous knowledge about the concerns in the sphere of the UFIL. These actors could come from economic development, health, tourism, planning, or a number of other government departments.
3. Stakeholder representatives: according to the relevance for each innovation along the food supply chain – producer, transformer, distributor, consumer, waste management, governance.
4. Sustainability experts: knowledgeable about different sustainability dimensions, not necessarily involved in innovation or in the URBAL team.

Both the practitioners who emerged in the field of interest, and the sustainability experts who have a more abstract view on linkages and results, can provide a well-balanced analysis. In reality, representatives might have multiple roles and responsibilities and can offer a range of diverse insights. Drawing a stakeholder affiliation map for each UFIL will be a valuable tool to identify the density of linkages and diversity of roles that are combined within one UFIL.

A logic model to help map impact pathways

The logic model used in our impact pathway mapping involves a representation of how sustainable innovation holders identify or construct one or more social concerns that they aim to address by formalizing an operational project (Figure 2.3). According to this project, inputs (financial, human, material, capital, and other resources that enable the intervention to generate a product or service) are defined and elaborated from the innovation context. Through innovation activities defined by the operational mission, actions transform inputs into outputs. As previously said, for URBAL, especially during the workshops, short-term, medium-term, and long-term 'results' are used instead of 'outcomes' and 'impacts' to account for the differences brought about by the innovation. Outcomes can be regarded as direct short-term effects on immediate stakeholders, while impacts refer to medium- to long-term effects that go beyond the scope of the direct actors and users. From a pragmatic point of view, this understanding sets the URBAL methodology apart from other methodology that focuses on academic purposes. Referring to 'outcomes', 'impacts' could lead to confusion as different disciplines and practitioner groups may have conflicting views on this term. 'Changes' and 'results' are simpler terms that facilitate work in a participative way. For example, it may not be appropriate to use the term 'impact', because most of the social innovations we will work on have been founded recently. To us, both 'changes' and 'results' are appropriate to compensate for the relative difficulty of 'impacts' as well as the uncertainty related to their identification.

The diagram presented here has been shaped at the very beginning to the URBAL project, before any live-test with an innovation. Since then, we have been able to perform this test twice and have significantly modified the methodology, from the logic model to the very practical aspects of its implementation. One example is the end of the distinction between outputs and results in favour of a more general distinction between short-term, medium-term, and long-term changes.

Twelve Urban Food Innovation Labs

The methodology will be tested on a total of 12 UFILs (Table 2.1) are engaged during the three years of the project in eight cities, including Montpellier (2), Milan (3), Hanoi (2), Rabat, Brasilia, Baltimore, Cape Town, and Berlin.

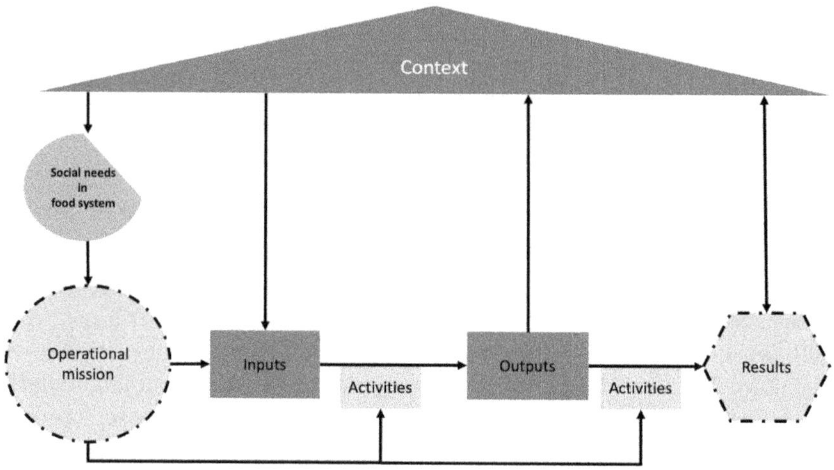

Figure 2.3 A frame for impact pathway mapping.
Source: URBAL.

- **Context(s)** of urban-driven social innovation (SI) for sustainable food systems consists of internal and external opportunities and obstacles. The context influences, and is influenced by, the SI.
  - Example: Economic, social, environmental, political, and cultural factors (at a local and at larger scale, depending on the relevance for each UFIL).
- **Sustainability needs in the food system** refers to demands in relation to food that are not yet being met. Such needs can actually exist or be only perceived by SI holders.
  - Example: Food poverty in an urban neighbourhood. While needs tend to be social in nature, they can be driven by other needs, for example, to improve soil quality.
- **Operational mission** describes the vision and strategies that SI holders set for their intervention to address a need. This mission is the guiding principle of the intervention and is the basis for input acquisition and activities to metabolize inputs into outputs. It roughly resembles the theory of change.
  - Example: Reducing food poverty through voluntary social work and support by the municipality to empower people to gain ownership of their own food supply.
- **Inputs** are the physical and non-physical factors needed to achieve the SI's objective. Inputs are retrieved from both the internal and external contexts.
  - Example: Volunteers, social innovation grants, and support from civil servants.
- **Activities** defines actions that are undertaken to use and transform inputs into outputs or generate results.
  - Example: Board meetings, job creation, teaching, sales, and applications.
- **Outputs** are all products and services directly generated by an SI through the transformation of inputs.
  - Examples: Mobile apps, cooking workshops, and food store concepts.
- **Results or changes** refer to changes that can be linked to an SI. These results are relative, that is, there is a noteworthy difference between the status quo and the previous condition. A result is the consequence of the application or use of an SI's output by the target group and other stakeholders. A result can also occur indirectly and unplanned, and alter conditions positively or negatively.
  - Example: Knowledge about the closest location to acquire healthy food.

UFILs were selected using a multifaceted typology based on the following four key questions: (a) Where are the changes expected: on consumer practices; on value-chains; and/or on governance of urban food policies? (b) What is the type of sustainability innovation, including the satisfaction of human needs, changes to social relations, increasing levels of socio-political capability, and asset building at individual and/or community levels (Kirwan et al., 2013)? (c) Who originates the innovation: civil society; the private sector; and/or local public authorities? (d) How many dimensions of sustainability are addressed among environmental, socio-cultural, economic, food security and nutrition, and governance? Only those innovations whose constructed vision involves at least two sustainability dimensions have been selected for participation.

Other criteria include: legitimacy and precedence of participative approaches; institutional context including whether food policies exist and if they do whether they are local and if they are led as top-down, bottom-up, or a mixture of initiatives; development and cultural contexts; stage of the innovation, from initial planning through emergent to fully developed and mature innovations; expected outcomes of the innovations, including scaling up, out, and deep (Riddell & Moore, 2015); and, status of the innovation as novel, adapted from elsewhere, or imitated.

Table 2.1 Innovation origins and expected changes for the 12 URBAL Urban Food Innovation Labs (UFILs)

	Changes in consumer practices	*Changes in value-chain organization*	*Changes in urban food policy governance*
Innovation from civil society	Collaborative consumer supermarket, **Montpellier**	Participatory guarantee system **Rabat**	Food Aid **Milan**
Innovation from private sector	Promotion of native species from the Cerrado biome in the gastronomy **Brasilia** Internet usage in Food provisioning and information **Hanoi**	Aquaponics as innovative supply chain **Berlin**	Community Food Committees for urban health and nutrition **Cape Town**
Innovation from public authorities	Support to innovations in school canteens, **Montpellier**	Long-distance contracts between city and hinterland **Hanoi** Public procurement for school canteens, **Milan**	Food Policy Advisor Network, **Baltimore** Distretto Agricolo Milanese **Milan**

The URBAL method will be adapted to the needs and context of each UFIL, thanks to a step-by-step improvement of the method (Figure 2.3). Live tests of the methodology will be done in three waves with feedback analysis from participants after each wave, which will help evolve the methodology. As at April 2019, two tests have already been performed, ten more are still to be organized. The final version of the methodology will be available at the conclusion of this phase of the project, projected for the end of 2020.

In each of these cities where innovations have been chosen, the main partners of the projects have strong links with scientific teams. Several of these cities are actively part of national or international networks on urban food systems. This will help to better connect the URBAL project to other cities and to disseminate the results of the project.

Conclusion

Testing this methodology on various innovations in diverse contexts will allow the researchers to present an analysis of the actual effects, the potentialities, the risks, and the limits of urban innovations on the sustainability of food systems (including outside the urban perimeter). In such a way, this research will contribute to the analysis of how city regions can be part of solutions and contribute to more sustainable food systems even if they are, to a large extent, responsible for the existing pressures on planetary boundaries.

The research outcomes are expected to demonstrate conditions when urban-driven initiatives, including urban food policies, may provide beneficial medium-term or long-term changes (i.e. impacts), helping urban areas to collaborate with rural partners in their regions in the transition towards more sustainable food systems. It is also expected that the lessons learned will in turn pose new challenges. One purpose of the research project is to contribute to reversing the image of cities as a main source of unsustainability as cities are frequently and deservedly associated with negative environmental impacts, unequal availability, and accessibility to balanced and affordable nutritious food for city dwellers. URBAL will explore urban spaces as sources of innovative solutions to sustainability issues. This, in turn, can be linked to other initiatives including the Milan Urban Food Policy Pact and strategies being developed to link the Sustainable Development Goals (SDGs) to urban food system practice (Calori et al., 2017; Blay-Palmer et al., 2018).

The approach also makes an important contribution at the intersection of impact pathways mapping literature and participatory methods by developing a more precise set of terms to describe the innovation process. This enables a more robust and complex analysis and responds to the needs and questions of the various stakeholders actually or potentially engaged in the innovation (Fawcett et al., 2003; Douthwaite et al., 2007a; Neubert, 2010; see also Chapter 1, this volume). In turn, it is expected that the specific innovations in the UFILs will allow the project to address questions about forces both external to the innovation process, including procedural routines, or the innovation aims, including marketing and funding goals (Dhondt et al., 2016, p. 22), or internal factors such as project management and learning and can provide specific information to foster learning, consensus building, and decision-making (Douthwaite et al., 2007b; Berg et al., 2009; Suárez-Herrera et al., 2009). That URBAL is inclusive and flexible, allows it to meet these goals as well as build capacity and agency for organizational stakeholders (Springer-Heinze et al., 2003; Marra, 2015; see also Chapter 4, this volume). As it is developed in tandem with practitioners and reflects their everyday practice, it is more usable, flexible, and relevant (Rodríguez-Campos, 2017; Zukoski & Bosserman, 2017).

Beyond research interest, this method will also help practitioners and innovation stakeholders to capture the very effect of their actions and give them the resources to enable them to be more reflexive and strategic about their innovation pathway, for their organization, as well as for their communication with policymakers and investors. The impact pathways will extract details and interconnections about the process and innovation and will make impacts/results more transparent and more easily understood. This method can be a low-tech and low-cost tool for policymakers, innovators, and funders to understand more about existing and proposed urban food innovations.

Acknowledgements

URBAL (N° 1507-200 AF; N° FC 2015–2440, N° FDNC Engt 00063479) is supported under the 'Thought for Food' Initiative of Agropolis Fondation (through the 'Investissements d'avenir' programme with reference number ANR-10-LABX-0001-01'), Fondazione Cariplo and Daniel & Nina Carasso Foundation. The Social Sciences and Humanities Research Council of Canada supports the FLEdGE Partnership, a Canadian partner. We also recognize the valuable contribution of our community and academic partners, as well as the URBAL WP leaders and the Scientific Advisory Committee.

Note

1 Specifically in the 2010s, the development of the distinct impact pathways approach has skyrocketed. Many interesting approaches, mainly from agricultural research, contributed to the improvement and adaptation of concepts to a diverse range of interest fields. Researchers have proposed and developed different methodologies that assess, model, analyse, and evaluate the processes that (most likely) lead to a specific impact, such as PIPA, ImpresS, IMPRESA, and Syalinnov.

References

Abi-Nader, J., Ayson, A., Harris, K., Herrera, H., Eddins, D., Habib, D., & Villanueva, L. (2009) *Whole measures for community food systems: Values-based planning and evaluation*. Burlington, VT, Center for Whole Communities.

Aubin, J., Donnars, C., Supkova, M., Dorin, B., Gaillard, G., Garcia, F., … Rojot, C. (2013) A critical panorama of methods used to assess food sustainability. In: Esnouf, C., Russel, M., & Bricas, N. (Eds.) *Food system sustainability: Insights from duALIne*. Cambridge, Cambridge University Press, pp. 198–232.

Avelino, F. & Rotmans, J. (2009) Power in transition: An interdisciplinary framework to study power in relation to structural change. *European Journal of Social Theory*. 12(4), pp. 543–569. Available at: https://doi.org/10.1177/1368431009349830.

Barret, D., Blundo Canto, G., Dabat, M.-H., Devaux-Spatarakis, A., Faure, G., Hainzelin, E., … Triomphe, B. (2018) *ImpresS methodological guide. Methodological guide to ex post impact of agricultural research in developing countries*. Montpellier, France, Cirad. Available at: https://doi.org/10.19182/agritrop/00006.

Bennett, C.J. & Howlett, M. (1992) The lessons of learning: Reconciling theories of policy learning and policy change. *Policy Sciences*. 25(3), pp. 275–294.

Berg, C., Causeman, B., Guenther, D., Hochgesang, T., Müller, R., Neubert, S., & Quack, M. (2009) Landkarte Wirkungsanalyseverfahren (Arbeitskreis 'Evaluation von Entwicklungspolitik in der DeGEval'). Available at: https://www.researchgate.net/profile/Susanne_Neubert2/publication/238786352_Herausgeber_Arbeitskreis_Evaluation_von_Entwicklungspolitik_in_der_DeGEval/links/5594d80608ae5d8f392f82e4/Herausgeber-Arbeitskreis-Evaluation-von-Entwicklungspolitik-in-der-DeGEval.pdf.

Blay-Palmer, A., Sonnino, R., & Custot, J. (2016) A food politics of the possible? Growing sustainable food systems through networks of knowledge. *Agriculture and Human Values*. 33(1), pp. 27–43.

Blay-Palmer, A., Santini, G., Dubbeling, M., Renting, H., Taguchi, M., & Giordano, T. (2018) Validating the city region food system approach: Enacting inclusive, transformational city region food systems. *Sustainability*. 10(5), p. 1680.

Born, B. & Purcell, M. (2006) Avoiding the local trap: Scale and food systems in planning research. *Journal of Planning Education and Research*. 26(2), pp. 195–207.

Bozeman, B. & Sarewitz, D. (2011) Public value mapping and science policy evaluation. *Minerva*. 49(1), pp. 1–23. Available at: https://doi.org/10.1007/s11024-011-9161-7.

Brand, C. (2015) *Alimentation et métropolisation: repenser le territoire à l'aune d'une problématique vitale oubliée*. Thèse de doctorat en géographie, université Grenoble Alpes, Laboratoire PACTE, Grenoble.

Brand, C., Bricas, N., Conaré, D., Daviron, B., Debru, J., Michel, L., & Soulard, C.-T. (Eds.) (2017) *Construire des politiques alimentaires urbaines*. Paris, Editions Quae.

Bricas, N. (2017) Les enjeux de l'urbanisation pour la durabilité des systèmes alimentaires. In: Brand, C. et al. (Eds.) *Construire des politiques alimentaires urbaines*. Concepts et démarches. Paris, Editions Quae, pp. 19–38.

Cabannes, Y. (2004) Participatory budgeting: A significant contribution to participatory democracy. *Environment and Urbanization*. 16(1), pp. 27–46.

Callon, M., Barthe, Y., & Lascoumes, P. (2001). *Agir dans un monde incertain*. Essai sur la démocratie technique, Seuil.

Calori, A., Dansero, E., Pettenati, G., & Toldo, A. (2017) Urban food planning in Italian cities: A comparative analysis of the cases of Milan and Turin. *Agroecology and Sustainable Food Systems*. 41(8), pp. 1026–1046.

Carlsson, L., Callaghan, E., Morley, A., & Broman, G. (2017) Food system sustainability across scales: A proposed local-to-global approach to community planning and assessment. *Sustainability*. 9(6), p. 1061.

Clancy, K. (1994) Commentary social justice and sustainable agriculture: Moving beyond theory. *Agriculture and Human Values*. 11(4), pp. 77–83.

Cousins, J.B. & Earl, L.M. (1992) The case for participatory evaluation. *Educational Evaluation and Policy Analysis*. 14(4), pp. 397–418.

Desrochers, P. & Shimizu, H. (2008) *Yes, we have no bananas: A critique of the 'food miles' perspective*. Fairfax, VA, Mercatus Center, George Mason University. Available at: https://www.mercatus.org/system/files/Yes_We_Have_No_Bananas__A_Critique_of_the_Food_Mile_Perspective.pdf.

De Zeeuw, H. & Dubbeling, M. (2015) Process and tools for multistakeholder planning of urban agro-food systems. In: de Zeeuw, H. & Drechsel, P. (Eds.) *Cities and Agriculture – Developing Resilient Urban Food Systems*. London, Earthscan Food and Agriculture, Routledge, pp. 56–87.

Dhondt, S., Van de Ven, H., Ziauberyte, R., Van der Torre, W., Cressey, P., Kaderabkova, A., … Terstriep, J. (2016) *Ex-ante impact assessment & value network analysis for SI*. Gelsenkirchen, Institute for Work and Technology. Available at: https://repository.tudelft.nl/view/tno/uuid%3A843a0c44-4e02-4213-9ff5-b03b0ff081d6.

Douthwaite, B., Alvarez, B.S., Cook, S., Davies, R., George, P., Howell, J., Mackay, R., & Rubiano, J. (2007b) Participatory impact pathways analysis: A practical application program theory in research for development. *The Canadian Journal of Program Evaluation*. 22(2), pp. 127–159.

Douthwaite, B., Schulz, S., Olanrewaju, A.S., & Ellis-Jones, J. (2007a) Impact pathway evaluation of an integrated Striga hermonthica control project in northern Nigeria. *Agricultural Systems*. 92(1–3), pp. 201–222. Available at: https://doi.org/10.1016/j.agsy.2006.03.007.

Ekirapa-Kiracho, E., Ghosh, U., Brahmachari, R., & Paina, L. (2017) Engaging stakeholders: Lessons from the use of participatory tools for improving maternal and child care health services. *Health Research Policy and Systems*. 15(Suppl 2), pp. 17–28. Available at: https://doi.org/10.1186/s12961-017-0271-z.

Fawcett, S.B., Boothroyd, R., Schultz, J.A., Francisco, V.T., Carson, V., & Bremby, R. (2003) Building capacity for participatory evaluation within community initiatives. *Journal of Prevention & Intervention in the Community*. 26(2), pp. 21–36. Available at: https://doi.org/10.1300/J005v26n02_03.

Fraser, D.G.E., Dougill, A.J., Mabee, W.E., Reed, M., & McAlpine, P. (2005) Bottom up and top down: analysis of participatory process for sustainability indicator identification as a pathway to community empowerment and sustainable environmental management. *Journal of Environmental Management*. 78, pp. 114–127.

Freidberg, S. (2003) Cleaning up down south: Supermarkets, ethical trade and African horticulture. *Social and Cultural Geography*. 4(1), pp. 27–43.

Friedmann, H. (2007) Scaling up: Bringing public institutions and food service corporations into the project for a local, sustainable food system in Ontario. *Agriculture and Human Values*. 24, pp. 389–398.

GIZ (Ed.) (2012) Wirkungsanalyse und Wirkungsmessung in Gesundheitsvorhaben der deutschen Entwicklungszusammenarbeit: Verbesserungsbedarf vor dem

Hintergrund internationaler Erfahrungen und Entwicklungen. Deutsche Gesellschaft für Internationale Zusammenarbeit (GIZ) GmbH. Available at: https://repository.publisso.de/resource/frl:5202307-1/data.

Hall, P.A. (1993) Policy paradigms, social learning, and the state: The case of economic policymaking in Britain. *Comparative Politics*. 25(3), pp. 275–296.

Howarth, R.B. & Norgaard R.B. (2013) Intergenerational Transfers and the Social Discount Rate. *Environmental and Resource Economics*. 3(4), pp. 337–358.

Hummelbrunner, R. (2007) Systemic evaluation in the field of regional development. In: Williams, B. & Imam, I. (Eds.) *Systems concepts in evaluation: An expert anthology*. Point Reyes, CA, EdgePress of Inverness for the American Evaluation Association, pp. 161–180.

Institute of Medicine and National Research Council. (2015) *A framework for assessing effects of the food system*. Washington, DC, The National Academies Press. https://doi.org/10.17226/18846

Jobert, B. & Muller, P. (1987) *L'état en action*. Politiques publiques et corporatismes, Paris, PUF.

Joly, P.B., Gaunand, A., Colinet, L., Larédo, P., Lemarié, S., & Matt, M. (2015) ASIRPA: A comprehensive theory-based approach to assessing the societal impacts of a research organization. *Research Evaluation*. 24(4), pp. 440–453.

Katz, D. (2010). The myth of food miles. In: *Fraser Forum* (April), pp. 21–23. Fraser Institute. Available at: https://www.fraserinstitute.org/sites/default/files/from-the-archive-myth-of-food-miles_0.pdf.

Kirwan, J., Ilbery, B., Maye, D., & Carey, J. (2013) Grassroots social innovations and food localization: An investigation of the local food programme in England. *Global Environmental Change*. 23(5), pp. 830–837. Available at: https://doi.org/10.1016/j.gloenvcha.2012.12.004.

Laidlaw, J. (2015) *Food security: A core component of a leading city's transformation agenda*. Melbourne, Australia, UN Global Compact Cities Programme.

Landert, J., Schader, C., Moschitz, H., & Stolze, M. (2017) A holistic sustainability assessment method for urban food system governance. *Sustainability*. 9(4), p. 490. Available at: https://doi.org/10.3390/su9040490.

Lang, T. (1999) The complexities of globalization: The UK as a case study of tensions within the food system and the challenge to food policy. *Agriculture and Human Values*. 16, pp. 169–185.

Leach, M., Scoones, I., & Stirling, A. (2010) *Dynamic sustainabilities: Technology, environment, social justice*. Washington, DC, Earthscan.

Luederitz, C., Schäpke, N., Wiek, A., Lang, D.J., Bergmann, M., Bos, J.J., ... Westley, F.R. (2017) Learning through evaluation: A tentative evaluative scheme for sustainability transition experiments. *Journal of Cleaner Production*. 169, pp. 61–76. Available at: https://doi.org/10.1002/9781119171386.ch3.

Mah, C.L. & Thang, H. (2013) Cultivating food connections: The Toronto food strategy and municipal deliberation on food. *International Planning Studies*. 18(1), pp. 96–110.

Marra, M. (2015) Cooperating for a more egalitarian society: Complexity theory to evaluate gender equity. *Evaluation*. 21(1), pp. 32–46. Available at: https://doi.org/10.1177/1356389014564856.

Mason, J.B. (1985) *Données requises pour l'évaluation des effets nutritionnels des projets de développement agricole et rural: document destiné aux planificateurs des projets*. Rome, FAO.

Mayne, J. (2011) Contribution analysis: Addressing cause and effect. In: Forss, K., Marra, M., & Schwartz, R. (Eds.) *Evaluating the complex: Attribution, contribution, and beyond*. Piscataway, NJ, Transaction Press.

McLaughlin, J.A. & Jordan, G.B. (2015) Using logic models. In: Newcomer, K.E., Hatry, H.P., & Wholey, J.S. (Eds.) *Handbook of practical program evaluation*. Hoboken, NJ, John Wiley & Sons, pp. 62–87.

Meter, K.A. (2007) Evaluating farm and food systems in the US. In: Williams, B. & Imam, I. (Eds.) *Systems concepts in evaluation: An expert anthology*. Point Reyes, CA, EdgePress of Inverness for the American Evaluation Association, pp. 141–159.

Moragues-Faus, A. & Marsden, T. (2017) The political ecology of food: Carving 'spaces of possibility' in a new research agenda. *Journal of Rural Studies*. *55*, pp. 275–288.

Naudet, J., Delarue, J., & Bernard, T. (2012) Impact evaluations: A tool for accountability? Lessons from experience in AFD. *Revue d'économie du développement*. 20(4), pp. 27–48. Available at: doi:10.3917/edd.264.0027.

Nelson, E., Tovar, L.G., Rindermann, R.S., & Cruz, M.Á.G. (2010) Participatory organic certification in Mexico: An alternative approach to maintaining the integrity of the organic label. *Agriculture and Human Values*. 27(2), pp. 227–237.

Neubert, S. (2010) *Description and examples of MAPP method for impact assessment of programmes and projects*. German Development Institute. Available at: http://www.ngo-ideas.net/mediaCache/MAPP/MAPP-Description.pdf.

OECD (2013) *Development results: An overview of results measurement and management*. Available at: https://www.oecd.org/dac/peer-reviews/Development-Results-Note.pdf.

Patton, M.Q. (2012) *Essentials of utilization-focused evaluation*. Thousand Oaks, CA, Sage Publications.

Pawson, R., Greenhalgh, T., Harvey, G., & Walshe, K. (2005) Realist review: A new method of systematic review designed for complex policy interventions. *Journal of Health Services Research & Policy*. 10(1), pp. 21–34. Available at: https://doi.org/10.1258/1355819054308530.

Peterson, M. (2006) The precautionary principle is incoherent. *Risk Analysis*. 26(3), pp. 595–601. Available at: https://doi.org/10.1111/j.1539-6924.2006.00781.x.

Plottu, B. & Plottu, É. (2009) Contraintes et vertus de l'évaluation participative. *Revue française de gestion*. 2(192), pp. 31–58.

Popov, E.V., Veretennikova, A.Y., & Omonov, Z.K. (2017) A social innovation impact assessment matrix. *Digest Finance*. 22(4), pp. 365–378. https://doi.org/10.24891/re.15.9.1752

Portney, P.R. & Weyant, J.P. (1999) *Discounting and intergenerational equity*. Baltimore, MD, Taylor & Francis.

Proietti, C., Wittine, D., Alvarez, S., Parker, M., Schulte-Geldermann, E., Stoian, D., & Thiele, G. (2015) *Co-constructing impact pathways with stakeholders for results-based management*. CGIAR Research Program on Roots, Tubers and Banana, Brief n°1.

Quiédeville, S., Barjolle, D., Mouret, J.C., & Stolze, M. (2017) Ex-post evaluation of the impacts of the science-based research and innovation program: A new method applied in the case of farmers' transition to organic production in the Camargue. *Journal of Innovation Economics*. 22(1), p. 145. Available at: https://doi.org/10.3917/jie.022.0145.

Quiédeville, S., Barjolle, D., Stolze, M., & Mouret, J.C. (2015) Impacts of science-based research and innovation program: The case of farmers' transition to organic production in Camargue. Conference paper at: *Impacts of agricultural research: an approach of societal values*. Paris, 3–4 November.

Reade, N. (2008) Konzept für alltagstaugliche Wirkungsevaluierungen in Anlehnung an Rigorous Impact Evaluations: Erprobung der Durchführung im Rahmen von GTZ Unabhängigen Evaluierungen. Available at: https://docplayer.org/55849-Konzept-fuer-alltagstaugliche-wirkungsevaluierungen-in-anlehnung-an-rigorous-impact-evaluations.html.

Reardon, T. (2015) The hidden middle: The quiet revolution in the midstream of agrifood value chains in developing countries. *Oxford Review of Economic Policy*. 31(1), pp. 45–63.

Reynolds, B. (2009) Feeding a world city: The London food strategy. *International Planning Studies*. 14(4), pp. 417–424.

Riddell, D. & Moore, M-L. (2015) Scaling out, Scaling up, Scaling deep: Advancing systemic social innovation and the learning processes to support it. Paper prepared for the J.W. McConnell Family Foundation and Tamarack Institute, Toronto and Waterloo, ON.

Robineau, O. (2015) Toward a systemic analysis of city-agriculture interactions in west Africa: A geography of arrangements between actors. *Land Use Policy*. 49, pp. 322–331.

Rocha, C. & Lessa, I. (2009) Urban governance for food security: The alternative food system in Belo Horizonte, Brazil. *International Planning Studies*. 14(4), pp. 389–400.

Rodríguez-Campos, L. (2017) Essentials of collaborative evaluation. In: Fetterman, D.M., Rodríguez-Campos, L., & Zukoski, A.P. (Eds.) *Collaborative, participatory, and empowerment evaluation: Stakeholder involvement approaches*. New York, Guilford Publications.

Rose, R. (1993) *Lesson-drawing in public policy: A guide to learning across time and space*. Chatham, NJ, Chatham House Publishers.

Sanyé-Mengual, E., Orsini, F., & Gianquinto, G. (2018) Revisiting the sustainability concept of urban food production from a stakeholders' perspective. *Sustainability*. 10(7), 2175. Available at: https://doi.org/10.3390/su10072175.

Singh, R.K., Murty, H.R., Gupta, S.K., & Dikshit, A.K. (2012) An overview of sustainability assessment methodologies. *Ecological Indicators*. 15(1), pp. 281–299.

Slocum, R., Cadieux, K., & Blumberg, R. (2016) Solidarity, space, and race: Toward geographies of agrifood justice. *Justice spatiale – Spatial Justice*. 9(January).

Springer-Heinze, A., Hartwich, F., Henderson, J.S., Horton, D., & Minde, I. (2003) Impact pathway analysis: An approach to strengthening the impact orientation of agricultural research. *Agricultural Systems*. 78(2), pp. 267–285. Available at: https://doi.org/10.1016/S0308-521X(03)00129-X.

Suárez-Herrera, J.C., Springett, J., & Kagan, C. (2009) Critical connections between participatory evaluation, organizational learning and intentional change in pluralistic organizations. *Evaluation*. 15(3), pp. 321–342. Available at: https://doi.org/10.1177/1356389009105884.

Temple, L., Barret, D., Blundo Canto, G., Dabat, M.H., Devaux-Spatarakis, A., Faure, G., & Triomphe, B. (2018) Assessing impacts of agricultural research for development: A systemic model focusing on outcomes. *Research Evaluation*. 27(2), pp. 157–170. Available at: https://doi.org/10.1093/reseval/rvy005.

UNEP (2011) *Decoupling natural resource use and environmental impacts from economic growth*. A report of the working group on decoupling to the international resource panel. United Nations Environment Programme, Nairobi, Kenya.

Viljoen, A. & Wiskerke, J.S. (2012) *Sustainable food planning: Evolving theory and practice*. Wageningen, Wageningen Academic Publishers.

Wertheim-Heck, S., Spaargaren, G., & Vellema, S. (2014) Food safety in everyday life: Shopping for vegetables in a rural city in Vietnam. *Journal of Rural Studies*. 35, pp. 37–48.

Wiek, A., Kay, B., & Forrest, N. (2017) Worth the trouble?! An evaluative scheme for urban sustainability transition labs (USTLs) and an application to the USTL in Phoenix, Arizona. In: De Niki, F., Loorbach, D., Castán Broto, V., & Coenen, L. (Eds.) *Urban sustainability transitions*. New York, Taylor & Francis, pp. 227–256.

World Commission on Environment and Development (1987) *Our common future*. Oxford, Oxford University Press.

Zukoski, A.P. & Bosserman, K. (2017) Essentials of participatory evaluation. In: Fetterman, D.M., Rodríguez-Campos, L., & Zukoski, A.P. (Eds.) *Collaborative, participatory, and empowerment evaluation: Stakeholder involvement approaches*. New York, Guilford Publications.

3 Taking care of the land

An interdisciplinary approach to community-based food systems assessment in Kakisa, Northwest Territories, Canada

Andrew Spring, Kelly Skinner, Melaine Simba, Erin Nelson, Jennifer Baltzer, Heidi Swanson, and Merritt Turetsky

Introduction

Food insecurity in Canada's north[1] has received national and international attention in recent years. In many northern communities, 17 per cent to 69 per cent of households report moderate to high food insecurity, which is dramatically higher than the national average of 8 per cent (Rosol et al., 2011; Council of Canadian Academies, 2014; Tarasuk et al., 2016). This discrepancy highlights the failure, at multiple scales, to ensure the availability, accessibility, and adequacy of food in northern communities. The right to food was guaranteed through Canada's commitments to the Universal Declaration of Human Rights and the International Covenant on Economic, Social, and Cultural Rights, but the lack of policy restricts enforcement of these rights (De Schutter, 2012). For communities in Canada's north, the right to food is more complex as issues of remoteness and lack of economic opportunities and infrastructure contribute to high rates of food insecurity. These issues are further complicated by external pressures of development, politics, and globalization (Power, 2008; Loring & Gerlach, 2009, 2015; Blay-Palmer et al., 2014; Council of Canadian Academies, 2014; see also Chapters 4 & 10, this volume). Factors contributing to food insecurity in Canada's north are numerous, complex, and subject to history, scale, and place. Redefining the discussion of food security, ensuring the socio-cultural aspects of place are included, while empowering communities to define their own food system, is critical (see Chapters 1 & 12, this volume). However, many of the essential ingredients that have been a part of northern communities' food systems for generations are poorly understood and not well represented in current measurements, metrics, or literature.

Food systems are defined by place and local circumstances (Marsden, 2012; Blay-Palmer et al., 2015; see also Chapter 6, this volume). In the Northwest Territories (NWT), the boreal forest is the most widespread ecosystem and Indigenous communities rely on this ecosystem to support their way of life. Traditional foods,[2] obtained though harvesting – hunting, fishing, and gathering – from the land and waters form the basis of the communities' food systems. These food systems have been sustainable for generations, and have adapted to changes in seasons, migrations, and availability of animals. It is a close connection and relationship with the land, and an understanding of natural variability that has enabled communities to thrive. The passing down of traditional knowledge through generations allows for youth to continue to access the land and provide food for the community. Being on, and living off, the land also includes elements of self-governance, practicing cultural and spiritual traditions, the social network, and support of the community (Bartlett, 2005; Parlee et al., 2007), as all are essential parts of the food system.

Food systems across the North have been, and continue to be, impacted by issues of colonialism, including forced placement in settlements, and other government policies, infrastructure, and the wage-based economy. As reliance on the formal economy increases, less time spent on the land has led to a decreased dependence on harvesting food, furs, and other products. Economic development, land rights, and management and protection of natural systems are all part of the complex situation that has developed in the North and have direct impacts on food security. The need for wage-based income is now crucial due to the high cost of living in the North, but employment opportunities are still sparse and leave less time to participate in traditional activities. With less time spent on the land harvesting traditional food, communities have become more dependent on market foods. As a consequence, diets of Indigenous communities in the NWT, consistent with changes across the globe, have moved away from traditional food sources to food purchased from stores (Kuhnlein & Receveur, 1996, 2007; Drewnowski & Popkin, 1997; Popkin, 2002; Kuhnlein et al., 2004; Damman et al., 2008). While the high cost of food and lack of affordable, nutritious options are often cited as major barriers to food security in the North, this does not tell the full story for this complex food system.

Currently, there is a lack of tools to assess food systems in the northern context and no existing tool that captures the complexity of northern food systems (see Chapters 4, 5, 9, 10, & 11, this volume). In part, this challenge stems from the existence of a mixed economy where traditional food systems and market food systems operate in parallel to supply sustenance, and other benefits to well-being, for people living in the North. The metrics that are available primarily focus on measuring the market food system. For example, the National Nutritious Food Basket (NNFB) and the Revised Northern Food Basket (RNFB) are standard tools accepted by statisticians, governments, and academics, among others, to monitor the price of food purchased from retailers. The NNFB and RNFB were created as survey instruments to measure the cost of a basic diet that met current nutrition recommendations and reflected average consumer purchasing patterns (INAC, 2007; Ministry of Health Promotion, 2010). The NNFB was created by Health Canada and the RNFB by Indigenous and Northern Affairs Canada specifically to monitor the cost of food in remote northern communities. The NNFB and RNFB each include their own unique list of 67 standardized food items and their purchase size (Ministry of Health Promotion, 2010). These food costing tools serve as one method to estimate much more complex food practices in households. They are not meant as a weekly shopping list or household budgeting tool; do not reflect food preferences; assume ability within the household to prepare meals from scratch; and do not include pre-packaged food, food eaten at restaurants, or locally sourced foods. As such, this metric does not capture the complexity of food systems in Canada's north, but only adds to the false argument that food security in the North is a purely economic and market-based issue.

We have little knowledge of, nor the infrastructure to track, traditional food procurement by communities, nor do we know what contribution traditional food has on local economies or livelihoods. Metrics for costing traditional food procurement have received even less research attention and there is no definitive methodology for costing harvested food. There are a few examples of discrete studies that have tried to estimate these costs. Harvester surveys have been used to quantify the amount of traditional food harvested by communities and also can monitor the number of harvesters. These tools can be applied to measure changes over time, which can have both policy and co-management implications for food species (Usher & Wenzel, 1987; Berkes, 1990; Wenzel et al., 2016). For example, Pal and colleagues (2013) asked active harvesters in Wapekeka and Kasabonika First Nations in northern Ontario to keep detailed logs of harvesting costs and recorded yields of hunts. They estimated the average cost of traditional meat was $14 per kilogram and that the annual cost for an active hunter was approximately $25,000. This assessment method is intensive and cannot be generalized to other harvesters or to communities in other northern regions. The 'Paying for Nutrition' project used the RNFB to assess market food costs in remote communities in northern Ontario (Veeraraghavan et al., 2016), and to acknowledge the mixed economy and the importance of traditional food systems (Skinner et al., 2016). The research team had many discussions about how to measure the costs of harvested foods and decided to use a proxy measure of hunting costs by gathering data on a short list of harvesting equipment (i.e., snare wire, fishing net, fishing line, and shotgun ammunition) as well as the cost of gasoline in the participating communities (Skinner et al., 2016). These items only consider some of the more frequent costs of going out on the land and water for food, but not some of the larger (and more expensive) equipment that might be required, such as a shotgun, boat and motor, all-terrain vehicle, or snowmobile. The amount of time spent on the land during harvesting should also be factored in, as it is time away from participating in the wage economy. Most importantly, there needs to be an acknowledgement that harvesting is not always fruitful, and hunters and harvesters may return home with no food. Combined, high costs associated with being on the land and the decline of harvesters has contributed to issues of food insecurity across the North where some cannot afford food either from the land or the store (Lambden et al., 2006; Ford et al., 2010; Ford & Beaumier, 2011; Brinkman et al., 2014; Loring & Gerlach, 2015). Additionally, the items that would need to be included in a more broadly relevant traditional food-costing assessment tool for the North would need to be appropriate and reflective of the diversity of local harvesting practices.

While the high cost of market food and other items in the North has been a major focus of a small number of previous studies, current food costing methods in northern environments have not considered the informal or social food economy. Food sharing is widely practised in northern environments and is an integral part of Indigenous culture and traditions (Condon et al., 1995; Collings et al., 1998; Chabot, 2003; Abele, 2009; Collings, 2011; Dombrowski et al., 2013; Skinner et al., 2013). Typically, harvesters return to the community and share food with immediate family members and community Elders, and sometimes these sharing networks extend to more distant kin and friends, including intercommunity connections. Communities therefore share the burden of costs and labour of harvesting by lending equipment, providing supplies, or assisting in meat processing. As selling of traditional food is not permitted in many communities (small commercial fisheries excluded) based on treaty or comprehensive land claim agreements, or social and cultural norms, it is thus only permitted in some jurisdictions (e.g., Nunavut) and also happens informally (Chan et al., 2006; Damman et al., 2008; Gombay, 2009). Currently people in some communities sell traditional food through social media platforms. However, with no formal system to measure or collect information on how food is distributed through social and informal economies, a large portion of the food system remains unmeasured.

Finally, the health of the ecosystem and its relationship to food systems requires attention in food system assessment. As the traditional food system is closely tied to the health of the ecosystem, protecting the land, and therefore the food system, has been prioritized by many communities and has been the subject of numerous natural science studies throughout the North. Research done in partnership with the community of Kakisa, NWT has combined natural sciences and community-based research with a food system approach. This intersecting research has enabled the community to lead and build a vision of a food system that relies on maintaining a healthy ecosystem in the face of climate change, growing foods, and developing opportunities for food sharing and economic benefits. As such, Kakisa offers valuable insight in rethinking how multifaceted food system evaluation should be done in Canada's north.

Community-driven food system metrics

The community of Kakisa is located approximately 400 km from Yellowknife, the capital of the NWT, Canada. Home to the Ka'a'gee Tu First Nation (KTFN), this community has a population of approximately 50 people, and is so small that it is often excluded from most maps of the region. The KTFN's traditional territory occupies approximately 10,000 km^2 within the northern boreal forest and is comprised of a mixture of forest, peatlands, wetlands, rivers, and lakes. The area includes two large lakes, Kakisa and Tathlina, which are connected by a series of rivers to the Mackenzie River. Community members enjoy a close relationship with the land and use the area for a wide variety of harvesting purposes including hunting, fishing, and gathering, all of which form a crucial component of their food system. The community's most important food source is fish, but moose and other large animals, such as woodland caribou, are harvested. Kakisa is a migratory stop for many species of waterfowl, which are typically hunted in the spring. As the community is small, it lacks a great deal of infrastructure and therefore depends on services offered in surrounding communities. The nearest grocery store is located approximately 150 km away, therefore additional resources, time, and money are required to access these services. Importantly, food from the store is perceived to be unhealthy (Spring et al., 2018).

Kakisa, like many communities in the North, faces several complex issues that impact the health of their food system, including climate change, development, and other political and socio-economic factors. This community is actively partnering in many research initiatives to help to understand these impacts and maintain their livelihoods for future generations. One such project was a climate change adaptation initiative, initiated by community members in 2014, and detailed in Spring et al. (2018). Through a participatory action research approach (see Chapters 2, 6, & 10, this volume), community members were able to identify barriers to accessing traditional foods and opportunities to overcome those barriers and build a food system that fitted their needs and vision. Much of the progress that has happened since 2014 has been through partnerships between the authors and community members working towards this vision. With gardens being installed, on-the-land camps that focus on youth–Elder knowledge transfer, a waste management programme, as well as different research and monitoring initiatives, much of the work has focused on protecting the land, improving access to fresh, healthy foods (both locally grown and traditional foods) and sharing knowledge with youth, Elders, and others around traditional skills and stewardship of the land. Reflecting on this work, the authors and community members, wanted to evaluate these past experiences as part of a process to assess the impact of these projects on the community's food system. In the following section, we examine aspects of the food system that the community is actively evaluating through research partnerships. This includes evaluation of the health of the ecosystem, and how that informs the basis of other food systems research in the community.

Health of the ecosystem

A healthy food system in a changing environment

In Kakisa, as well as all other northern communities, the connection to the forest is a critical part of the food system as a direct source of food. However, in food system evaluation discourse where agriculture is the dominant food source, the natural environment is considered to play a small role in the overall health of the food system. This disconnect is problematic for understanding and evaluating northern food systems. For members of the KTFN, the health of the natural environment is closely linked to the health of the people (Spring et al., 2018). While there is no formal assessment of the health of the KTFN Territory, researchers have been working closely with the community for years, in an endeavour to understand various components of the ecosystem, particularly in the context of anthropogenic climate change. This information is key to understanding the context for food system well-being through the lens of the changing climate.

Climate warming is amplified in the North; in the NWT, temperatures have increased approximately 3°C in the last 60 years, and more so during the winter months (Marshall & Baltzer, 2015), a rate that is 3–4 times the global average (IPCC, 2018). Responses of northern ecosystems to this change are already evident and are expected to become more severe in coming decades causing concern for the impacts on communities and their food systems. Rapid climate change in northern Canada has impacted and will continue to impact the health of northern ecosystems. Climate change is altering forest productivity directly while simultaneously accelerating large-scale disturbance processes including permafrost thaw, wildfire, and insect pest outbreaks (Gauthier et al., 2015). Overlain on these effects of climate change are the direct human impacts on forest health, including forest management decisions and the introduction of invasive species. Our knowledge of these issues remains incomplete in the region as these are novel or rapidly changing disturbances (Turetsky et al., 2017). However, for the community, these are pressing issues that need to be understood to support adaptation and planning for the protection of their livelihoods and food security for future generations. Here we provide a brief overview of these issues with a focus on impacts for Kakisa's community food system. These changing environmental conditions should be considered in the design of a healthy food system, ideally with long-term monitoring via community observations to determine their impacts on local food systems.

Disturbances influenced by climate change

Drought

Because the productivity of northern forests is thought to be limited by temperature, boreal productivity was predicted to increase with climate warming (Hogg & Bernier, 2005). However, growth responses of high latitude forests have been variable, with many forests showing reduced productivity in recent decades in response to warming temperatures (Walker & Johnstone, 2014). This has been largely attributed to decreasing water availability, which is expected to occur with warming despite increases in precipitation in many places (Price et al., 2013). This imbalance means that in many places trees are experiencing water limitation, particularly in northwestern North America (Hogg & Bernier, 2005). Tree growth also can decline with permafrost thaw (Patankar et al., 2015; Walker et al., 2015; Sniderhan & Baltzer, 2016). Permafrost containing ice can directly affect soil moisture through its control on the water table position.

Gradual versus abrupt thaw of permafrost

Permafrost thaw is one of the main challenges facing communities in the North. Approximately 70 per cent of the entire boreal forest is underlain by permafrost, a term given to soil, rocks, and sediment that remains at or below 0°C for two consecutive years. In the NWT, permafrost ranges from continuous coverage in the north where permafrost is thick, to discontinuous coverage of thinner permafrost further south. In part because of permafrost, the dominant tree species in the NWT rely on a shallow rooting zone to support growth. Warming leads to thickening of the active layer (seasonally thawed soil near the surface), a process that gradually affects surface permafrost over decades. In response to gradual permafrost thaw, plant productivity may be stimulated by a thicker rooting zone and increased nutrient availability. On the other hand, drying of surface soils may lead to declining tree growth (Walker et al., 2015; Sniderhan & Baltzer, 2016).

Where permafrost is ice-rich, it can thaw abruptly leading to a variety of surface expressions that include erosional features (retrogressive thaw slumps, active layer detachments, gullies), thaw lakes, or thaw wetlands. All of these surface expressions involve thermokarst (ground subsidence) and instability as a result of the abrupt thaw of ice-rich permafrost. The development of thermokarst features alter the local hydrology, ecology, and biogeochemistry (Jorgenson & Osterkamp, 2005; Quinton et al., 2011; Kokelj & Jorgenson, 2013; Baltzer et al., 2014; Olefeldt et al., 2016). Rapid rates of climate warming have already pushed parts of the southern NWT across critical thresholds for maintaining permafrost. For example, the southern limit of discontinuous permafrost was shown to have migrated northward by approximately 120 km between 1964 and 1989 (Kwong & Gan, 1994), likely transitioning KTFN territory from extensive discontinuous to sporadic discontinuous permafrost.

The region surrounding Kakisa is dominated by low-lying and poorly drained land with a mosaic of permafrost-free wetlands and slightly elevated permafrost peatlands that are better drained and support black spruce-dominated forest (Baltzer et al., 2014). Abrupt thaw in these systems leads to surface subsidence and associated waterlogging of the soils, resulting in 'drunken trees' and rapid tree mortality associated with inundation (Baltzer et al., 2014; Patankar et al., 2015). This is representative of the forest-to-wetland conversion occurring across the southern NWT and also around the circumboreal region (Baltzer et al., 2014; Helbig et al., 2016; Turetsky et al., 2019). This fundamental change in land cover and ecology leads to issues of food access and availability for communities as well as concerns about harvester safety. Large-scale conversion of forested ecosystems to wetlands will decrease habitat for some important species, including caribou, through loss of spruce-lichen forests (Joly et al., 2012). Abrupt thaw and thermokarst also results in biogeochemical changes that can affect the safety of food resources, for example mobilization of mercury (St. Pierre et al., 2018). Finally, changes in topography can make it more difficult for harvesters to access the land and uneven terrain can contribute to safety concerns for land users (Kokelj & Jorgenson, 2013; IPCC, 2014).

In response to all of these landscape changes in their region, the community has responded by developing an innovative web map, referred to as the Ka'a'gee Tu Atlas, in partnership with researchers. This Atlas was constructed for the community, collecting data layers from various sources that contained information on traditional land uses, culturally significant places as well as environmental information. Through focus groups, interviews, and modified photovoice techniques community members have compiled images and descriptions of what the land looked like before many of the impacts of climate change. This initiative has created a baseline for future monitoring initiatives that will record how the land is changing and provide the opportunity for community members to record current conditions, hazards, to share and discuss areas of concern on the land that may require monitoring or future research, and track impacts to their food system.

Wildfire

Fire is a critical part of the natural renewal of the boreal forest. However, climate warming is accelerating the frequency and intensity of wildfires. In 2014, the NWT experienced an historically unprecedented fire season, both in terms of area burned (2.85 million hectares; Walker et al., 2018) and impacts on communities (Dodd et al., 2018). Wildfires also have an immediate impact on wildlife habitat, as large areas are cleared during a fire, however patterns of regeneration of forest following wildfire has meant that these impacts are temporary. Many communities, including Kakisa, are concerned with whether the current impacts of climate change will allow the forest to recover to its original state. Severe wildfires, in both Alaska and the Yukon, have demonstrated the potential for fast-growing, deciduous species to become dominant after an intense fire where combustion of the organic soil layer is complete (Johnstone et al., 2010b). This can occur in response to very severe (deep burning) ground fires or those returning within short time intervals and can impact tree and ground vegetation recovery (Johnstone et al., 2010a; Hollingsworth et al., 2013). Understanding the nature of this shift is important with respect to the availability of wildlife habitat under an accelerated fire regime.

To answer these questions researchers established a network of permanent sampling plots in fire scars from 2014. These plots are unique in that they provide a co-located set of measures of fire severity (canopy and soil), permafrost conditions, pre-fire stand age and structure, detailed characterization of post-fire soil conditions (residual organic matter, soil carbon content and loss, bulk density, and nutrient status), and post-fire vegetation recovery, which will provide novel insights into the impacts of the 2014 fires. Some of these locations were on the traditional lands of the KTFN, which was important as the community was evacuated, and nearly lost, during the 2014 fires. Researchers helped to facilitate on-the-land learning events with the community to provide a demonstration on how the impacts of forest fire are measured, and how forest regeneration is monitored. Community Elders participated in identifying plants, sharing the Dene language names and traditional uses with youth and researchers, and providing insights into what animals might come to the areas. Together, this research and community experience with fire can help inform and educate other regions impacted by forest fire along with future research directions in affected communities.

Fish, water, pests, and contaminants

One of the challenges facing a warming north is the loss of temperature limitation on the distribution of species that have previously been unable to expand their ranges into these regions. The introduction of new species can have detrimental impacts on the balance already present in the ecosystem (Walther et al., 2002). Furthermore, new species have the potential to bring diseases or parasites into new systems and impact the health of wildlife populations. Ticks and other diseases causing death in a variety of animal species have been recorded and will likely continue to spread across the north (Kutz et al., 2009). Changes to species' migratory patterns, populations, and health will ultimately have negative impacts on the community as well. If country food is no longer readily available, harvesters will have to invest more time and money into the pursuit of food, increasing the cost to the harvester. In Kakisa, a range of new species has appeared in past years. Deer have been spotted in the southern parts of their territory (Spring et al., 2018). There is evidence of cougars in the area. Most concerning for the community is the presence of pelicans on both Tathlina and Kakisa lakes due to the potential impact of this new species on fish populations on which the community depends both for commercial and subsistence harvesting.

Contamination of country food species, particularly fish, is a major concern for communities in the North. Pollutants, mainly produced at lower latitude, are transported northward and can be deposited into marine or terrestrial environments where they can build up in food webs and reach high concentrations in top predator species. At high enough concentrations, human consumers of fish can face health risks due to exposure from contaminants. Detectable levels of mercury have been recorded in various fish species from lakes around the NWT (Lockhart et al., 2005), and levels in several lakes and fish species have led to development of site-specific consumption advisories. As climate changes, there is an increased risk to communities as pathways of contamination, both biological (species and processes) and physical (transportation of contaminants) intensify (Kraemer et al., 2005; Stern et al., 2012). Although restrictions on chemical applications and controls on global emissions of chemicals in other regions of the globe may help lower future risks of contamination, there is a concern that contaminants locked away in frozen sediment could be made biologically available due to climate change (Loseto et al., 2004; Stern et al., 2012) and permafrost thaw can release stored mercury into downstream aquatic systems (Klaminder et al., 2008). The issues of contamination are critical to the food security dialog in the North. Quantifying both risks (e.g., contaminant concentrations) and benefits (e.g., nutritional quality) of traditional foods, and how these risks and benefits may be altered with climate change, is thus of paramount importance.

The KTFN has been involved in monitoring the health of waters around the community as well as fish populations for years as food fished from Tathlina and Kakisa lakes are of significant commercial and subsistence importance to the community. Mercury has been recorded in different species in both lakes, including walleye and northern pike (Laird et al., 2018). Some samples have been measured at levels that often exceed Health Canada guidelines for safe consumption. Researchers are looking to find the cause behind these elevated levels by increasing fish sample sizes in both lakes, augmenting sampling and the understanding of mercury accumulation in lower trophic levels in the food web (e.g., algae and plankton) and determining the amount of mercury in sediment that is available for uptake into the food web. Furthermore, research has been ongoing to understand the levels of contaminants in community members. This would help to better understand community fish consumption habits and lead to improved communication and warnings about certain fish species, sizes of fish, and community impacts of contaminants.

Supporting community participation in food system evaluation

One of the most important aspects of the research conducted in Kakisa is how it is driven by the community. The community, through the Band Council and the Environmental Coordinator, has a great deal of experience working with researchers from universities and government departments, as well as other organizations, to conduct a variety of research projects on their lands. The community has the vision for ecosystem and food system protection and works with researchers to achieve their goals of stewardship of their lands. For many of the researchers that work with the KTFN, the research began at the invitation of the community, either directly or through other direct relationships. Being a small community there is the understanding that there is a limited capacity to do research, and a great deal of questions and concerns about how climate change and other developments are impacting the ecosystem. Partnering with researchers therefore became the way for the community to learn more about these changes, build local capacity to participate in, and support research and benefit from, project funding and employment opportunities.

Another unique element of the research is the partnership, and personal relationships, built with the KTFN's Environmental Coordinator. This individual has been involved in environmental monitoring projects through government departments and non-governmental organizations, and has coordinated field sampling for fish, water, and wildlife monitoring and social science research in the community. This position is critical to ensuring that all research is conducted alongside representatives of the community (Berkes & Jolly, 2002; Pearce et al., 2009; McGregor et al., 2010; Armitage et al., 2011; Tondu et al., 2014). Of most benefit to the projects are the connections this role has in the small community that have allowed for easy communication and connections to community members, Elders, and local decision makers as well as fostered a bridge between all parties to build trust and transparent communication, so all research directly benefits the community (Angell & Parkins, 2010; Tondu et al., 2014). This role helps to overcome several major barriers for researchers working in the North, specifically the limited time one has to spend in the field, and builds upon the social capital of an individual or partner organization within the community (Caine et al., 2007). What emerged as a powerful driver behind the research and projects is the coordinator's personal drive to conduct work that makes a positive impact in the community. This drive not only enabled them to champion much of the research being conducted, but pushed much of the research, beyond mere participatory and collaborative, to action research.

The impact that community participation has on the research goes much further than logistics and relationship building, it ensures the research meets the needs of the community. This includes fostering a two-eyed seeing approach that emphasizes both traditional and Western knowledge, and research methodologies (Bartlett et al., 2012; Martin, 2012). The research being conducted with the community all builds towards the KTFN's vision to establish a legally protected conservation area in line with the environmental and cultural values of the community. The proposed Protected Area encompasses much of KTFN's traditional territory including valuable hunting and fishing areas used by community members. Part of the planned management of the area will be accomplished through the establishment of a local Indigenous Guardian programme to promote stewardship of the land and waters through improved environmental monitoring (Trant et al., 2012; Social Ventures Australia, 2016). Guardians will be community members who will act as the eyes and ears on the land as they patrol, monitor, report, and manage a range of research and monitoring activities. The establishment of local, regional and national Indigenous Guardian initiatives is an emerging effort to reclaim sovereignty over traditional lands and promote self-determination (Roburn, 2018). In the context of food systems, this is the realization of what food sovereignty looks like in the North, providing rights to lands, decision-making, management, and protection of ecosystems for future generations.

To achieve the goals of the Guardian programme, the community requires additional capacity in various areas to be fully and properly engaged in research and monitoring programmes. Capacity needs include training, equipment, infrastructure, administrative support, and other aspects of programme development and delivery. Assessing the training needs and building in links to the opportunities available to address those needs is another important aspect in maintaining the community's food system. For the KTFN, it is critically important to engage youth in research and give them the opportunity to learn all the skills they need to participate in monitoring the land, but also to survive and harvest food from the land. Youth are important stakeholders in the North, and opportunities to learn from both Elders and researchers while on the land shapes their abilities to be future leaders in their communities. Several on-the-land camp experiences have been facilitated in Kakisa and surrounding communities to give youth the opportunity to learn traditional skills, see the land from both the traditional knowledge and Western science perspectives, and encourage their involvement in environmental monitoring initiatives in their communities. Through on-the-land training in both scientific and traditional knowledge methods of studying environmental change, participating youth develop as leaders in food system stewardship and research. But as the opportunities for this training are typically funded through external programmes, we have limited knowledge about how to effectively engage youth in these skills, let alone continue to develop these skills over time. Schools are incorporating on-the-land learning into the curriculum and facilitating more programmes to be on the land, and learning traditional skills are critical to the long-term health of the food system and well-being of communities, and therefore becomes an important aspect of measuring the food system.

Local food production

Kakisa, like many other communities, has identified growing food as a key way forward to building a more resilient food system (Chen & Natcher, 2019; Spring et al., 2018; see also Chapter 9, this volume). Not only will local food production provide fresh, healthy food to Kakisa and potentially other nearby communities, it also avoids the high cost of purchasing food from the store and the environmental impacts of transportation to northern communities (Chen & Natcher, 2019). The story of growing food in Kakisa is long and influenced by both positive and negative experiences with government programmes, residential schools, and people who have come through the community in the past (Simba & Spring, 2017). Growing food in Kakisa most recently started in 2015 with the creation of small raised-bed gardens and has expanded over the years to include another small plot of land for potatoes. Recent conversations in the community have asked about the impact of the gardens, and it was noted that not enough food is being produced to make a significant impact on the community's food system. Although the gardens are a source of pride for community members, scaling up to small-scale agriculture is envisioned as the next step in building a more sustainable food system in the community. Across the North, but particularly in the southern part of the NWT where Kakisa is located, climate change causing warming temperatures and soils as well as faster nutrient cycling may support more productive growing environments for food crops. However, the potential for local food production in the NWT more broadly depends on many factors, such as long-term capacity to grow food, the support for community initiatives and policies, and environmental factors including soil fertility and whether those soils will become wetter or drier with climate warming.

For Kakisa, scaling up production of gardens, and expanding growing opportunities in the community's planned fire break are opportunities to grow enough food to feed the community, and likely create economic opportunities for some community members. For larger communities, however, more land will need to be developed for agriculture to meet the needs of the population, and potentially, when policy barriers are mitigated and distribution networks are created, feed other communities in the NWT and beyond. There is potential for agricultural expansion into the boreal forest to have negative impacts on the ecosystem and potentially release carbon into the atmosphere to further contribute to climate change. Thus, the impacts of local food production on the health of the northern food system will need to be assessed as will methods of growing food that may reduce some of these potential impacts.

If local food production is to increase in the future, it is important to identify the areas that will become more suitable for agriculture, build capacity to grow food, and develop policies and practices that help protect the land, soils, and atmosphere (see Chapter 4, this volume). The northern food system and agroecology share many commonalities, but the deeply place-based nature of traditional and agroecological food systems means that experiences from other jurisdictions cannot be directly imported into

Canada's NWT. That said, knowledge generated through international agroecology research and practice still holds relevance for conversations about how to move forward in communities such as Kakisa. Specifically, insights can be gleaned from examining how regions and communities with long histories of engagement with agroecology are assessing the functioning of their food systems.

Agroecology has been defined as the ecology of the food system (Francis et al., 2003; Gliessman, 2007) and an agroecological lens requires attention to be paid to all elements of that system. This includes the complex web that exists amongst food production, procurement, processing, transportation, consumption, and waste management, as well as the interrelated elements of ecological soundness, economic viability, and social justice (Gliessman, 2007, 2013). Francis et al. (2003, p. 103) argue that agroecology compels us 'to evaluate food systems in new ways, to recognize the need to balance the system with available resources, and to accept a moral obligation to manage outputs from the system in an equitable manner.' While this complex, whole systems approach renders development and application of simple assessment tools difficult, if not impossible, efforts have been made to explore the links between agroecology and various measures of a food system's sustainability. One of the important contributions of the agroecological approach to food system assessment is that it explicitly centres on ecosystem health, thus addressing a gap in many of the more commonly applied assessment methods that focus more heavily on socio-economic metrics. Many ecosystem elements have been considered in agroecology-based impact assessments, including specific metrics of soil health and water quality (Altieri et al., 2017), and Francis et al. (2003) suggest employing a range of tools, including life cycle analysis (Clift et al., 1997) and environmental footprint calculations (Wackernagel et al., 1999). In recent discourse, two of the most important components that have emerged as central to agroecological approaches to food system assessment are biodiversity and climate change resilience.

Conserving biodiversity – particularly that of locally adapted species – is a cornerstone of agroecological practice (Gliessman, 2007; Altieri & Nicholls, 2012). As a result, rather than taking individual species yields as a measure of success, agroecology-based assessment seeks to capture the synergistic benefits associated with more diversified food ecologies (Badgley et al., 2007; Altieri & Nicholls, 2012). In Cuba, a recognized global leader in agroecology, Leyva and Lores (2018) have developed a biodiversity-based method for measuring agroecosystem sustainability that evaluates species populations destined for human and animal consumption, soil nourishment, and broader socio-ecological functions such as carbon sequestration or inclusion in spiritual ceremonies and traditions. This method involves communities establishing baselines for productive potential and comparing potential diversity against actual levels of diversity. One reason for using biodiversity as a metric, or even proxy, for food system sustainability is that more diversified systems have been found to be linked to increased resilience in the face of climate change (Holt-Giménez, 2002; Rosset, 2011;

Altieri et al., 2015). Much of this work on assessing agroecological systems in the context of a changing climate and threats to the world's biodiversity highlights the fact that

> farmers living in harsh environments in Africa, Asia, and Latin America have developed and/or inherited complex farming systems managed in ingenious ways. These systems have allowed small farming families to meet their subsistence needs in the midst of environmental variability without depending on modern agricultural technologies.
>
> (Altieri et al., 2015, p. 874)

The traditional knowledge held by food producers in these 'harsh environments' is essential to agroecological food system assessment, as is a focus on evaluating priorities determined through participatory, community-driven processes. As a result, many agroecology-based food system evaluations are grounded in participatory methodologies such as farmer-to-farmer learning (see Kangmennaang et al., 2017) or Participatory Rural Appraisal (Arnés et al., 2018). Such methods allow for holistic assessments that consider the inter-relationships between various socio-ecological metrics, including biodiversity, resilience to climate change, food security and sovereignty, energy efficiency, and community well-being, all of which are fundamental to an agroecological food system. Attention to the political realities within which food systems are located is also necessary, as policies at various scales have been shown to 'lock in' unsustainable practices, while also having the potential to incentivize healthier alternatives (IPES-Food, 2016).

For Kakisa, working with partners to learn more about agroecological methods fits within the community's desire to protect the land. As agriculture is, and will remain, a small portion of the community's overall food system, ensuring it does not compromise the health of the boreal forest is critical. In fact, learning how agroecology can be carried out in the NWT in a way that can help to build resilience into the boreal forest, sequester carbon and provide access to fresh, healthy food in a context specific to Kakisa is part of a larger collaboration led by the community over the next few years. This could shape policies that will ensure that agriculture is developed in a way that meets the needs of local residents and does not diminish the overall health of the ecosystem. Through learning from other regions, such as Cuba, and sharing experiences and knowledge with other communities, Kakisa can benefit from, but also enrich the global dialogue around food systems evaluation and research.

Conclusion

Food system evaluation and research in Canada's north is complex as factors that drive the health and sustainability of food systems are diverse, multidisciplinary, and place-based, but subject to issues of scale. For the most part, evaluation tools that exist do not capture the unique and complicated food system that continues to thrive in the NWT (see Chapters 2, 4, 5, & 9, this volume). In the small community of Kakisa, examples are shown of how a community is championing efforts to protect their food system for future generations by partnering on a variety of research initiatives with academic and government institutions and other organizations. The community's participation in research is critical not only to the success of these initiatives and capacity building for community-led research, but a part of their desire to enhance control over their own food system. The environmental and climate change research in and around Kakisa shows that food systems need to be better understood in holistic terms, and that the most important element is the health of the land. This is the guiding principle for all development, including agriculture, for the community. This sends an important message to those who conduct research in the North, that is, that all changes to the ecosystems impact community food systems, livelihoods, and well-being. Most, if not all, researchers are part of an ongoing evaluation of the health of food systems across the North, but research disciplines silo those results. As food systems researchers, we cannot be bound by research disciplines but need to work to connect disciplines. Research efforts like those in Kakisa strive to create multidisciplinary approaches to environmental challenges and their impacts on community livelihoods, but connecting those results and initiatives remains a work in progress. There are other examples to learn from (Wolfe et al., 2011; Crate et al., 2017), however all involve building meaningful relationships and fostering partnerships with the communities where the research takes place. Indigenous world views see these links across disciplines, and their involvement in research is critical for the future of their food systems. Therefore, supporting community-driven research and fostering stronger inclusion and communication of research processes on ecosystem change become a critical element of self-determination and food sovereignty.

Acknowledgements

The authors would like to acknowledge the Ka'a'gee Tu First Nation for their support and partnership as part of these research endeavours. This research was made possible by funding contributions from various federal and territorial agencies, and partnerships with multiple collaborators.

Notes

1 Canada's north can be defined as comprising the three Northern Territories of Yukon, Northwest Territories, Nunavut as well as Nunavik (northern Quebec) and Nunatsiavut (in Labrador). However, this definition may also include Indigenous communities from remote and Northern regions of the provinces, such as Ontario and Alaska, who rely on the boreal forest for their livelihoods.

2 The authors use the term 'traditional foods' to describe foods that are harvested or gathered from the land and waters, but literature and other sources will use terms like country or wild foods.

References

Abele, F. (2009) The state and the northern social economy: Research prospects. *The Northern Review*. 30(Spring), pp. 37–56.

Altieri, M., Nicholls, C., & Montalba, R. (2017) Technological approaches to sustainable agriculture at a crossroads: An agroecological perspective. *Sustainability*. 9, p. 349. Available at: https://doi.org/10.3390/su9030349.

Altieri, M.A. & Nicholls, C.I. (2012) Agroecology scaling up for food sovereignty and resiliency. In: Lichtfouse, E. (Ed.) *Sustainable agriculture reviews*. Dordrecht, Springer Netherlands, pp. 1–29. Available at: https://doi.org/10.1007/978-94-007-5449-2_1.

Altieri, M.A., Nicholls, C.I., Henao, A., & Lana, M.A. (2015) Agroecology and the design of climate change-resilient farming systems. *Agronomy for Sustainable Development*. 35(3), pp. 869–890. Available at: https://doi.org/10.1007/s13593-015-0285-2.

Angell, A.C. & Parkins, J.R. (2010) Resource development and aboriginal culture in the Canadian north. *Polar Record*. 47(1), pp. 67–79. Available at: https://doi.org/10.1017/S0032247410000124.

Armitage, D., Berkes, F., Dale, A., Kocho-schellenberg, E., & Patton, E. (2011) Co-management and the co-production of knowledge: Learning to adapt in Canada's arctic. *Global Environmental Change*. 21(3), pp. 995–1004. Available at: https://doi.org/10.1016/j.gloenvcha.2011.04.006.

Arnés, E., Astier, M., Marín González, O., & Hernández Díaz-Ambrona, C.G. (2018) Participatory evaluation of food and nutritional security through sustainability indicators in a highland peasant system in Guatemala. *Agroecology and Sustainable Food Systems*. 43(5), pp. 482–513. Available at: https://doi.org/10.1080/21683565.2018.1510871.

Badgley, C., Moghtader, J., Quintero, E., Zakem, E., Chappell, M.J., Avilés-Vázquez, K., Samulon, A., & Perfecto, I. (2007) Organic agriculture and the global food supply. *Renewable Agriculture and Food Systems*. 22(2), pp. 86–108. Available at: https://doi.org/10.1017/S1742170507001640.

Baltzer, J.L., Veness, T., Chasmer, L.E., Sniderhan, A., & Quinton, W.L. (2014) Forests on thawing permafrost: Fragmentation, edge effects, and net forest loss. *Global Change Biology*. 20(3), pp. 824–834. Available at: https://doi.org/10.1111/gcb.12349.

Bartlett, C., Marshall, M., & Marshall, A. (2012) Two-eyed seeing and other lessons learned within a co-learning journey of bringing together Indigenous and mainstream knowledges and ways of knowing. *Journal of Environmental Studies*. 2, pp. 331–340.

Bartlett, J.G. (2005) Health and well-being for Métis women in Manitoba. *Canadian Journal of Public Health*. 96, pp. S22–S27.

Berkes, F. (1990) Native subsistence fisheries: A synthesis of harvest studies in Canada. *ARCTIC*. 43(1), pp. 35–42. Available at: https://doi.org/10.14430/arctic1588.

Berkes, F. & Jolly, D. (2002) Adapting to climate change: Social-ecological resilience in a Canadian western arctic community. *Conservation Ecology*. 5(2), p. 18.

Blay-Palmer, A., Knezevic, I., & Spring, A. (2014) Seeking common ground for food system transformation. *Dialogues in Human Geography*. 4(2), pp. 185–189. Available at: https://doi.org/10.1177/2043820614537154.

Blay-Palmer, A., Sonnino, R., & Custot, J. (2015) A food politics of the possible? Growing sustainable food systems through networks of knowledge. *Agriculture and Human Values*. 33(1), pp. 27–43. Available at: https://doi.org/10.1007/s10460-015-9592-0.

Brinkman, T., Maracle, K.B., Kelly, J., Vandyke, M., Firmin, A., & Springsteen, A. (2014) Impact of fuel costs on high-latitude subsistence activities. *Ecology and Society*. 19(4), p. 27. Available at: https://doi.org/10.5751/ES-06861-190418.

Caine, K.J., Salomons, M.J., & Simmons, D. (2007) Partnerships for social change in the Canadian north: Revisiting the insider–outsider dialectic. *Development and Change*. 38(3), pp. 447–471.

Chabot, M. (2003) Economic changes, household strategies, and social relations of contemporary Nunavik Inuit. *Polar Record*. 39(1), pp. 19–34. Available at: https://doi.org/10.1017/S0032247402002711.

Chan, H.M., Fediuk, K., Hamilton, S., Rostas, L., Caughey, A., Kuhnlein, H., Egeland, G., & Loring, E. (2006) Food security in Nunavut, Canada: Barriers and recommendations. *International Journal of Circumpolar Health*. 65(5), pp. 416–431.

Chen, A. & Natcher, D. (2019) Greening Canada's arctic food system: Local food procurement strategies for combating food insecurity. *Canadian Food Studies*. 6(1), pp. 140–154.

Clift, R., Audsley, A., Alber, S., Cowell, S., Crettaz, P., Gaillard, G., Hausheer, J., Jolliet, O., Kleijn, R., Mortensen, B., et al. (1997) *Harmonisation of Environmental Life Cycle Assessment for Agriculture*. Surrey, Silsoe Research Institute.

Collings, P. (2011) Economic strategies, community, and food networks in Ulukhaktok, Northwest Territories, Canada. *Arctic*. 64(2), pp. 207–219.

Collings, P., Wenzel, G., & Condon, R.G. (1998) Modern food sharing networks and community integration in the central Canadian arctic. *Arctic*. 51(4), pp. 301–314.

Condon, R.G., Collings, P., & Wenzel, G. (1995) The best part of life: Subsistence and economic adaptation adult Inuit males among young. *Arctic*. 48(1), pp. 31–46.

Council of Canadian Academies (2014) *Aboriginal food security in northern Canada: An assessment of the state of knowledge*. Ottawa, ON, Council of Canadian Academies.

Crate, S., Ulrich, M., Habeck, J.O., Desyatkin, A.R., Desyatkin, R.V., Fedorov, A.N., Hiyama, T., Iijima, Y., Ksenofontov, S., Mészáros, C., & Takakura, H. (2017) Permafrost livelihoods: A transdisciplinary review and analysis of thermokarst-based systems of indigenous land use. *Anthropocene*. 18, pp. 89–104. Available at: https://doi.org/10.1016/j.ancene.2017.06.001.

Damman, S., Eide, W.B., & Kuhnlein, H.V. (2008) Indigenous peoples' nutrition transition in a right to food perspective. *Food Policy*. 33(2), pp. 135–155. Available at: https://doi.org/10.1016/j.foodpol.2007.08.002.

De Schutter, O. (2012) *Report of the special rapporteur on the right to food, mission to Canada*. New York, United Nations General Assembly.

Dodd, W., Scott, P., Howard, C., Scott, C., Rose, C., Cunsolo, A., & Orbinski, J. (2018) Lived experience of a record wildfire season in the Northwest Territories, Canada. *Canadian Journal of Public Health*. 109(3), pp. 327–337. Available at: https://doi.org/10.17269/s41997-018-0070-5.

Dombrowski, K., Khan, B., Channell, E., Moses, J., Mclean, K., & Misshula, E. (2013) Kinship, family, and exchange in a Labrador Inuit community. *Arctic Anthropology*. 50(1), pp. 89–104.

Drewnowski, A. & Popkin, B.M. (1997) The nutrition transition: New trends in the global diet. *Nutrition Reviews*. 55(2), pp. 31–43.

Ford, J.D. & Beaumier, M. (2011) Feeding the family during times of stress: Experience and determinants of food insecurity in an Inuit community. *The Geographical Journal*. 177(1), pp. 44–61. Available at: https://doi.org/10.1111/j.1475-4959.2010.00374.x.

Ford, J.D., Pearce, T., Duerden, F., Furgal, C., & Smit, B. (2010) Climate change policy responses for Canada's Inuit population: The importance of and opportunities for adaptation. *Global Environmental Change*. 20(1), pp. 177–191. Available at: https://doi.org/10.1016/j.gloenvcha.2009.10.008.

Francis, C., Lieblein, G., Gliessman, S., Breland, T.A., Creamer, N., Harwood, R., Salomonsson, L., Helenius, J., Rickerl, D., Salvador, R., Wiedenhoeft, M., Simmons, S., Allen, P., Altieri, M., Flora, C., & Poincelot, R. (2003) Agroecology: The ecology of food systems. *Journal of Sustainable Agriculture*. 22(10), pp. 99–118. Available at: https://doi.org/10.1300/J064v22n03_10.

Gauthier, S., Bernier, P., Kuuluvainen, T., Shvidenko, A.Z., & Schepaschenko, D.G. (2015) Boreal forest health and global change. *Science*. 349(6250), pp. 819–822. Available at: https://doi.org/10.1126/science.aaa9092.

Gliessman, S. (2013) Agroecology: Growing the roots of resistance. *Agroecology and Sustainable Food Systems*. 37(1), pp. 19–31. Available at: https://doi.org/10.1080/10440046.2012.736927.

Gliessman, S.R. (2007) *Agroecology: The ecology of sustainable food systems*. Second edition. Boca Raton, FL, CRC Press. Available at: https://doi.org/10.1201/b17420.

Gombay, N. (2009) Sharing or commoditising? A discussion of some of the socio-economic implications of Nunavik's Hunter Support Program. *Polar Record*. 45(2), p. 119. Available at: https://doi.org/10.1017/S003224740800778X.

Helbig, M., Pappas, C., & Sonnentag, O. (2016) Permafrost thaw and wildfire: Equally important drivers of boreal tree cover changes in the Taiga Plains, Canada: Drivers of boreal tree cover changes. *Geophysical Research Letters*. 43(4), pp. 1598–1606. Available at: https://doi.org/10.1002/2015GL067193.

Hogg, E.H., & Bernier, P.Y. (2005) Climate change impacts on drought-prone forests in western Canada. *The Forestry Chronicle*. 81(5), pp. 675–682. Available at: https://doi.org/10.5558/tfc81675-5.

Hollingsworth, T.N., Johnstone, J.F., Bernhardt, E.L., & Chapin, F.S. (2013) Fire severity filters regeneration traits to shape community assembly in Alaska's boreal forest. *PLoS ONE*. 8, (2), p. e56033. Available at: https://doi.org/10.1371/journal.pone.0056033.

Holt-Giménez, E. (2002) Measuring farmers' agroecological resistance after hurricane Mitch in Nicaragua: A case study in participatory, sustainable land management impact monitoring. *Agriculture, Ecosystems & Environment*. 93(1–3), pp. 87–105. Available at: https://doi.org/10.1016/S0167-8809(02)00006-3.

Indian and Northern Affairs Canada (INAC) (2007) *The revised northern food basket*. Ottawa, ON, Government of Canada.

IPCC (2014) *Climate change 2014: impacts, adaptation, and vulnerability. Part B: regional aspects*. Contribution of working group II to the Fifth Assessment Report of the Intergovernmental Panel on Climate Change. Cambridge, Cambridge University Press.

IPCC (2018) *Global warming of 1.5 °C*. An IPCC special report on the impacts of global warming of 1.5°C above pre-industrial levels and related global greenhouse gas emission pathways, in the context of strengthening the global response to the threat of climate change.

IPES-Food (2016) *From uniformity to diversity: a paradigm shift from industrial agriculture to diversified agroecological systems*. Louvain-la-Neue, Belgium, IPES.

Johnstone, J.F., Chapin, F.S., Hollingsworth, T.N., Mack, M.C., Romanovsky, V., & Turetsky, M. (2010a) Fire, climate change, and forest resilience in interior Alaska. *Canadian Journal of Forest Research*. 40(7), pp. 1302–1312. Available at: https://doi.org/10.1139/X10-061.

Johnstone, J.F., Hollingsworth, T.N., Chapin, F.S., & Mack, M.C. (2010b) Changes in fire regime break the legacy lock on successional trajectories in Alaskan boreal forest. *Global Change Biology*. 16(4), pp. 1281–1295. Available at: https://doi.org/10.1111/j.1365-2486.2009.02051.x.

Joly, K., Duffy, P., & Rupp, T. (2012) Simulating the effects of climate change on fire regimes in Arctic biomes: Implications for caribou and moose habitat. *Ecosphere*. 3(5), pp. 1–18.

Jorgenson, M.T. & Osterkamp, T.E. (2005) Response of boreal ecosystems to varying modes of permafrost degradation 1. *Canadian Journal of Forest Research*. 35(9), pp. 2100–2111. Available at: https://doi.org/10.1139/X05-153.

Kangmennaang, J., Kerr, R.B., Lupafya, E., Dakishoni, L., Katundu, M., & Luginaah, I. (2017) Impact of a participatory agroecological development project on household wealth and food security in Malawi. *Food Security*. 9(3), pp. 561–576. Available at: https://doi.org/10.1007/s12571-017-0669-z.

Klaminder, J., Yoo, K., Rydberg, J., & Giesler, R. (2008) An explorative study of mercury export from a thawing palsa mire. *Journal of Geophysical Research*. 113, G04034. Available at: https://doi.org/10.1029/2008JG000776.

Kokelj, S.V. & Jorgenson, M.T. (2013) Advances in thermokarst research. *Permafrost and Periglacial Processes*. 24(2), pp. 108–119. Available at: https://doi.org/10.1002/ppp.1779.

Kraemer, L.D., Berner, J.E., & Furgal, C.M. (2005) Climate change and human health the potential impact of climate on human exposure to contaminants in the Arctic. *International Journal of Circumpolar Health*. 64(5), pp. 498–508.

Kuhnlein, H.V. & Receveur, O. (1996) Dietary change and traditional food systems of indigenous peoples. *Annual Review of Nutrition*. 16, pp. 417–442. Available at: https://doi.org/10.1146/annurev.nu.16.070196.002221.

Kuhnlein, H.V. & Receveur, O. (2007) Local cultural animal food contributes high levels of nutrients for arctic Canadian Indigenous adults and children. *The Journal of Nutrition*. 137, pp. 1110–1114.

Kuhnlein, H.V, Receveur, O., Soueida, R., & Egeland, G.M. (2004) Arctic Indigenous peoples experience the nutrition transition with changing dietary patterns and obesity. *Journal of Nutrition*. 134(6), pp. 1447–1453.

Kutz, S.J., Jenkins, E.J., Veitch, A.M., Ducrocq, J., Polley, L., Elkin, B., & Lair, S. (2009) The arctic as a model for anticipating, preventing, and mitigating climate change impacts on host–parasite interactions. *Veterinary Parasitology*. 163(3), pp. 217–228. Available at: https://doi.org/10.1016/j.vetpar.2009.06.008.

Kwong, Y.T.J. & Gan, T.Y. (1994) Northward migration of permafrost along the Mackenzie highway and climatic warming. *Climatic Change*. 26(4), pp. 399–419. Available at: https://doi.org/10.1007/BF01094404.

Laird, M.J., Henao, J.J.A., Reyes, E.S., Stark, K.D., Low, G., Swanson, H.K., & Laird, B.D. (2018) Mercury and omega-3 fatty acid profiles in freshwater fish of the Dehcho Region, Northwest Territories: Informing risk–benefit assessments. *Science of the Total Environment*. 637, pp. 1508–1517. Available at: https://doi.org/10.1016/j.scitotenv.2018.04.381.

Lambden, J., Receveur, O., Marshall, J., & Kuhnlein, H. (2006) Traditional and market food access in arctic Canada is affected by economic factors. *International Journal of Circumpolar Health*. 65(4), pp. 331–340. Available at: https://doi.org/10.3402/ijch.v65i4.18117.

Leyva, Á. & Lores, A., 2018. Assessing agroecosystem sustainability in Cuba: A new agrobiodiversity index. *Elementa: Science of the Anthropocene*. 6, p. 80. Available at: https://doi.org/10.1525/elementa.336.

Lockhart, W.L., Stern, G.A., Low, G., Hendzel, M., Boila, G., Roach, P., Evans, M.S., Billeck, B.N., DeLaronde, J., Friesen, S., Kidd, K., Atkins, S., Muir, D.C.G., Stoddart, M., Stephens, G., Stephenson, S., Harbicht, S., Snowshoe, N., Grey, B., Thompson, S., & DeGraff, N. (2005) A history of total mercury in edible muscle of fish from lakes in northern Canada. *The Science of the Total Environment*. 351, pp. 427–463. Available at: https://doi.org/10.1016/j.scitotenv.2004.11.027.

Loring, P.A. & Gerlach, S.C. (2009) Food, culture, and human health in Alaska: An integrative health approach to food security. *Environmental Science & Policy*. 12(4), pp. 466–478. Available at: https://doi.org/10.1016/j.envsci.2008.10.006.

Loring, P.A. & Gerlach, S.C. (2015) Searching for progress on food security in the North American north: A research synthesis and meta-analysis of the peer-reviewed literature. *Arctic*. 68(3), p. 380. Available at: https://doi.org/10.14430/arctic4509.

Loseto, L.L., Siciliano, S.D., & Lean, D.R.S. (2004) Methylmercury production in high arctic wetlands. *Environmental Toxicology and Chemistry*. 23(1), pp. 17–23.

Marsden, T. (2012) Food systems under pressure: Regulatory instabilities and the challenge of sustainable development. In: Spaargaren, G., Oosterveer, P., & Loeber, A. (Eds.) *Food practices in transition: Changing food consumption, retail and production in the age of reflexive modernity*. New York, Routledge.

Marshall, K.E. & Baltzer, J.L. (2015) Decreased competitive interactions drive a reverse species richness latitudinal gradient in subarctic forests. *Ecology*. 96(2), pp. 461–470. Available at: https://doi.org/10.1890/14–0717.1.

Martin, D.H. (2012) Two-eyed seeing: A framework for understanding Indigenous and non-Indigenous approaches to Indigenous health research. *Canadian Journal of Nursing Research*. 44(2), p. 24.

McGregor, D., Bayha, W., & Simmons, D. (2010) 'Our responsibility to keep the land alive': Voices of northern Indigenous researchers. *Pimatisiwin: A Journal of Aboriginal and Indigenous Community Health*. 8(1), pp. 101–124.

Ministry of Health Promotion (2010) *Nutritious food basket guidance document*. Ottawa, ON, Government of Ontario, Standards, Programs & Community Development Branch.

Olefeldt, D., Goswami, S., Grosse, G., Hayes, D., Hugelius, G., Kuhry, P., McGuire, A.D., Romanovsky, V.E., Sannel, A.B.K., Schuur, E.A.G., & Turetsky, M.R.

(2016) Circumpolar distribution and carbon storage of thermokarst landscapes. *Nature Communications*. 7(1), p. 13043. Available at: https://doi.org/10.1038/ncomms13043.

Pal, S., Haman, F., & Robidoux, M.A. (2013) The costs of local food procurement in two northern Indigenous communities in Canada. *Food and Foodways*. 21(2), pp. 132–152. Available at: https://doi.org/10.1080/07409710.2013.792193.

Parlee, B., Neil, J.O., & Lutsel K'e Dene First Nation (2007) 'The Dene way of life': Perspectives on health from Canada's north. *Journal of Canadian Studies*. 41(3), pp. 112–133.

Patankar, R., Quinton, W.L., Hayashi, M., & Baltzer, J.L. (2015) Sap flow responses to seasonal thaw and permafrost degradation in a subarctic boreal peatland. *Trees*. 29(1), pp. 129–142. Available at: https://doi.org/10.1007/s00468-014-1097-8.

Pearce, T.D., Ford, J.D., Laidler, G.J., Smit, B., Duerden, F., Allarut, M., Andrachuk, M., Baryluk, S., Dialla, A., Elee, P., Goose, A., Ikummaq, T., Joamie, E., Kataoyak, F., Loring, E., Meakin, S., Nickels, S., Shappa, K., Shirley, J., & Wandel, J. (2009) Community collaboration and climate change research in the Canadian Arctic. *Polar Research*. 28(1), pp. 10–27. Available at: https://doi.org/10.1111/j.1751-8369.2008.00094.x.

Popkin, B.M. (2002) The shift in stages of the nutrition transition in the developing world differs from past experiences! *Public Health Nutrition*. 5(1a), pp. 205–214. Available at: https://doi.org/10.1079/PHN2001295.

Power, E.M. (2008) Conceptualizing food security for Aboriginal people in Canada. *Canadian Journal of Public Health*. 99, pp. 95–97.

Price, D.T., Alfaro, R.I., Brown, K.J., Flannigan, M.D., Fleming, R.A., Hogg, E.H., Girardin, M.P., Lakusta, T., Johnston, M., Mckenney, D.W., Pedlar, J.H., Stratton, T., Sturrock, R.N., Thompson, I.D., Trofymow, J.A., & Venier, L.A. (2013) Anticipating the consequences of climate change for Canada's boreal forest ecosystems. *Environmental Review*. 21(4), pp. 322–365.

Quinton, W.L., Hayashi, M., & Chasmer, L.E. (2011) Permafrost-thaw-induced land-cover change in the Canadian subarctic: Implications for water resources. *Hydrological Processes*. 25, pp. 152–158.

Roburn, S. (2018) Power from the north: The energized trajectory of Indigenous sovereignty movements. *Canadian Journal of Communication*. 43, pp. 167–184. Available at: https://doi.org/10.22230/cjc.2018v43n1a3310.

Rosol, R., Huet, C., Wood, M., Lennie, C., Osborne, G., & Egeland, G.M. (2011) Prevalence of affirmative responses to questions of food insecurity: International Polar Year Inuit health survey, 2007–2008. *International Journal of Circumpolar Health*. 70(5), pp. 488–497. Available at: https://doi.org/10.3402/ijch.v70i5.17862.

Rosset, P. (2011) Food sovereignty and alternative paradigms to confront land grabbing and the food and climate crises. *Development*. 54(1), pp. 21–30. Available at: https://doi.org/10.1057/dev.2010.102.

Simba, M. & Spring, A. (2017) Growing a garden in Kakisa. *Northern Public Affairs*. 5(April), pp. 24–26.

Skinner, K., Burnett, K., Williams, P., Martin, D., Stothart, C., LeBlanc, J., Veeraraghavan, G., & Sheedy, A. (2016) Challenges in assessing food environments in northern and remote communities in Canada. *Canadian Journal of Public Health*. 107, p. 60. Available at: https://doi.org/10.17269/cjph.107.5324.

Skinner, K., Hanning, R.M., Desjardins, E., & Tsuji, L.J.S. (2013) Giving voice to food insecurity in a remote indigenous community in subarctic Ontario, Canada: Traditional ways, ways to cope, ways forward. *BMC Public Health*. 13(1), p. 427. Available at: https://doi.org/10.1186/1471-2458-13-427.

Sniderhan, A.E. & Baltzer, J.L. (2016) Growth dynamics of black spruce (Picea mariana) in a rapidly thawing discontinuous permafrost peatland: Growth dynamics boreal peatlands. *Journal of Geophysical Research: Biogeosciences*. 121(12), pp. 2988–3000. Available at: https://doi.org/10.1002/2016JG003528.

Social Ventures Australia (2016) *Analysis of the current and future value of Indigenous Guardian work in Canada's Northwest Territories*. Sydney, Australia, Social Ventures Australia.

Spring, A., Carter, B., & Blay-Palmer, A. (2018) Climate change, community capitals, and food security: Building a more sustainable food system in a northern Canadian boreal community. *Canadian Food Studies*. 5, pp. 111–141.

St. Pierre, K.A., Zolkos, S., Shakil, S., Tank, S.E., St. Louis, V.L., & Kokelj, S.V. (2018) Unprecedented increases in total and methyl mercury concentrations downstream of retrogressive thaw slumps in the western Canadian arctic. *Environmental Science and Technology*. 52(24), pp. 14099–14109. Available at: https://doi.org/10.1021/acs.est.8b05348.

Stern, G.A., Macdonald, R.W., Outridge, P.M., Wilson, S., Chételat, J., Cole, A., Hintelmann, H., Loseto, L.L., Steffen, A., Wang, F., & Zdanowicz, C. (2012) How does climate change influence Arctic mercury? *Science of the Total Environment*. 414, pp. 22–42. Available at: https://doi.org/10.1016/j.scitotenv.2011.10.039.

Tarasuk, V., Mitchell, A., & Dachner, N. (2016) *Household food insecurity in Canada, 2014*. Toronto, ON, Research to identify policy options to reduce food insecurity (PROOF).

Tondu, J.M.E., Balasubramaniam, A.M., Chavarie, L., Gantner, N., Knopp, J.A., Provencher, J.F., Wong, P.B.Y., & Simmons, D. (2014) Working with northern communities to build collaborative research partnerships: Perspectives from early career researchers. *Arctic*. 67(3), pp. 419–429.

Trant, A.J., Jacobs, J.D., & Sable, T. (2012) Teaching and learning about climate change with Innu Environmental Guardians. *Polar Geography*. 35(3–4), pp. 229–244. Available at: https://doi.org/10.1080/1088937X.2012.682229.

Turetsky, M.R., Abbott, B.W., Jones, M.C., Walter Anthony, K., Olefeldt, D., Schuur, E.A.G., Koven, C., McGuire, A.D., Grosse, G., Kuhry, P., Hugelius, G., Lawrence, D.M., Gibson, C., & Sannel, A.B.K. (2019) Permafrost collapse is accelerating carbon release. *Nature*. 569(7754), pp. 32–34. Available at: https://doi.org/10.1038/d41586-019-01313-4.

Turetsky, M.R., Baltzer, J.L., Johnstone, J.F., Mack, M.C., McCann, K., & Schuur, E.A.G. (2017) Losing legacies, ecological release, and transient responses: Key challenges for the future of northern ecosystem science. *Ecosystems*. 20(1), pp. 23–30. Available at: https://doi.org/10.1007/s10021-016-0055-2.

Usher, P.J. & Wenzel, G. (1987) Native harvest surveys and statistics: A critique of their construction and use. *Arctic*. 40(2), pp. 145–160. Available at: https://doi.org/10.14430/arctic1759.

Veeraraghavan, G., Burnett, K., Skinner, K., Williams, P., Martin, D., Jamal, A., Ramsay, M., & Stothart, C. (2016) *Paying for nutrition: A report on food costing in the North*. Montreal, QC, Food Secure Canada.

Wackernagel, M., Onisto, L., Bello, P., Callejas Linares, A., Susana López Falfán, I., Méndez García, J., Isabel Suárez Guerrero, A., & Guadalupe Suárez Guerrero, M. (1999) National natural capital accounting with the ecological footprint concept. *Ecological Economics*. 29(3), pp. 375–390. Available at: https://doi.org/10.1016/S0921-8009(98)90063-5.

Walker, X. & Johnstone, J.F. (2014) Widespread negative correlations between black spruce growth and temperature across topographic moisture gradients in the boreal forest. *Environmental Research Letters*. 9(6), 064016. Available at: https://doi.org/10.1088/1748-9326/9/6/064016.

Walker, X.J., Mack, M.C., & Johnstone, J.F. (2015) Stable carbon isotope analysis reveals widespread drought stress in boreal black spruce forests. *Global Change Biology*. 21(8), pp. 3102–3113. Available at: https://doi.org/10.1111/gcb.12893.

Walker, X.J., Rogers, B.M., Baltzer, J.L., Cumming, S.G., Day, N.J., Goetz, S.J., Johnstone, J.F., Schuur, E.A.G., Turetsky, M.R., & Mack, M.C. (2018) Cross-scale controls on carbon emissions from boreal forest megafires. *Global Change Biology*. 24(9), pp. 4251–4265. Available at: https://doi.org/10.1111/gcb.14287.

Walther, G.R., Post, E., Convey, P., Menzel, A., Parmesan, C., Beebee, T.J.C., Fromentin, J.M., Hoegh-Guldberg, O., & Bairlein, F. (2002) Ecological responses to recent climate change. *Nature*. 416(6879), pp. 389–395. Available at: https://doi.org/10.1038/416389a.

Wenzel, G.W., Dolan, J., & Brown, C. (2016) Wild resources, harvest data and food security in Nunavut's Qikiqtaaluk region. *Arctic*. 69(2), pp. 147–159. Available at: https://doi.org/10.14430/arctic4562.

Wolfe, B.B., Humphries, M.M., Pisaric, M.F.J., Balasubramaniam, A.M., Burn, C.R., Chan, L., Cooley, D., Froese, D.G., Graupe, S., Hall, R.I., Lantz, T., Porter, T.J., Roy-Leveillee, P., Turner, K.W., Wesche, S.D., & Williams, M. (2011) Environmental change and traditional use of the Old Crow Flats in northern Canada: an IPY opportunity to meet the challenges of the new northern research paradigm. *Arctic*. 64(1), p. 127. Available at: https://doi.org/10.14430/arctic4092.

4 Assessing food systems as complex adaptive systems

Conceptual views and US applications

Ken Meter

Using a complexity approach can simplify

Scholars have long noted that paying close attention to complexity can simplify both assessment (Holling, 2001) and implementation (Holland, 2006). Four decades of professional practice through a non-profit research and consulting firm have confirmed this. Our partners are often community-based food system leaders who seek us out because, as they engage in systemic change efforts, they find that conditions are so complicated, and the multitude of options open to them so vast, they are unsure how to prioritize action.

Engage residents as systems experts

The central role of the food system assessor in this setting centres upon *engaging local residents as experts on local systems*, learning how they view the assets and weaknesses of their community, and understanding the potential solutions that are either emerging, or being stifled. This places the researcher in the position of *co-constructing* knowledge and potential policy frames (Blay-Palmer et al., 2018; Chapter 12, this volume). As an analysis is formed, the process aims to help build the capacity of local food system leaders to identify key systems levers, and to make strategic plans to move them during implementation. Developing systemic theories of change, local leaders can define indicators of progress that cut across issues that are often separated by disciplinary divides.

To date, this approach has been adopted by 140 regions in 40 states, 2 provinces, and by 4 Native American tribes (Crossroads Resource Center, n.d.). In each case, time-series economic data from public sources were used to illuminate changing dynamics in the local context. Where resources allowed a more in-depth approach (n=24), close interviews with wise practitioners helped uncover core system dynamics. Two examples later in this chapter will illustrate this approach.

Viewing food systems as complex adaptive systems (CAS)

Researchers often attempt to break down food systems into their component parts for analysis (Folke et al., 2005; Aubin et al., 2013; Koopmans 2017).

Yet, this ultimately proves difficult to defend. Folke et al. (2005) argue that the goal of adaptive management should not be to seek 'detailed knowledge of parts of the system, but improved understanding of the dynamics of the whole system' (p. 445) (Chapter 2, this volume). Aubin et al. (2013) caution that static indicators must be combined with more dynamic measures. As he analyses four different approaches to complexity, Koopmans (2017) notes that in certain contexts it is more important to analyse the *potential for change* in a system, rather than analysing the system itself in meticulous detail.

Moreover, focusing solely on a given food system is incomplete. Morin (2008) finds that *any* system co-produces itself in relation to its external environment, drawing energy from that environment, and manifesting autonomy and emergence within that context. Moreover, he adds that any definition of a system is inherently a mental abstraction created by an observer, and thus intimately connected to the world view of that observer. Morin (1992) further suggests that the term 'system' is generative to new ways of thinking that essentially lead to a '*paradigm* of complexity' (p. 6, emphasis added): a fundamental approach that calls for the observer to self-critically reflect on her knowledge as she engages a system (Chapters 2, 3, 5, 6, 7, 8, 9, 10, & 12, this volume). This requires 'thinking that is capable of unifying concepts which repel each other and are otherwise catalogued and isolated in separate compartments' (Morin, 2008, p. 81). He considers the paradigm of complexity to be 'a bold challenge to the fragmentary and reductionist spirit that continues to dominate the scientific enterprise' (Morin, 1992, p. 1). He adds that '[c]omplex thinking is not omniscient thinking ... It is always local, situated in a given time and place' (Morin, 1992, p. 97).

Food systems are constantly evolving in response to changing demands by society (Collona et al., 2013). Holling (2001) defines the term 'sustainability' as 'the capacity to create, test, and maintain adaptive capacity' (p. 390). This, of course, implies that our approach to assessing sustainable food systems must itself be adaptive.

While several scholars have approached food systems as complex (Stevenson et al., 2007; Aubin et al., 2013; Prosperi, 2015; Allen & Prosperi, 2016; Allen et al., 2018; Blay-Palmer et al., 2018), fewer have examined complex food systems as adaptive (Heffernan, 1999; Meter, 2007b, 2010; Swanson & Bhadwal, 2009; Nelson & Stroink, 2013; Stroink & Nelson, 2013; Prosperi, 2016).

Defining complex adaptive systems

One pioneer in the complexity field, John Holland (2006), defines complex adaptive systems (CAS) as 'systems that have a large number of components that adapt or learn as they interact' (p. 1). Holland (1992) adds, '[t]hese systems change and reorganize their component... parts to adapt themselves to the problems posed by their surroundings... From the standpoint of those who wish to understand and control these systems they constitute a moving target' (p. 18). Aubin et al. (2013) point out that static indicators 'do not take into account the adaptation and resilience of the systems studied' (p. 216).

This has profound impacts for analysts. The measurements one takes on a given food system at time *a* may be outmoded by time *b*, just a few weeks or months later. Indeed, the fundamental dynamics may have shifted even before measurements are compiled (Holland, 1992; Flood, 1999; Chapter 12, this volume). The basic nature of the entity may have changed so much that the measurements hold limited meaning, as explained below.

Consider the 2012 Census of Agriculture (US National Agricultural Statistical Service, series), which assembled data on US farming during one of its economic peaks. While census takers were gathering their tallies, grain prices were abnormally high (see Figure 4.1), fuelled by speculators who invested in commodity markets after suffering losses during the global housing finance crisis, as well as by strong demand for corn ethanol. This, in turn, meant that meat prices rose, since livestock are fed grain. By the time the 2012 counts were released in early 2014, conditions had changed substantially. The speculative bubble had burst. While corn producers in the US Midwest had earned a profit of US$129 per acre (€44 per ha) in 2012, they experienced a net loss of US$93 per acre (€32 per ha) in 2014 (ERS, Corn Production Costs and Returns, series). This is not to say the 2012 data were invalid. It does suggest that interpretation must take the changing nature of the system into account – in this case how a commodity economy that had brought rewards to producers rapidly shifted to one that created losses.

Food systems are especially prone to change in this emergent era. New industries have cropped up, for example, delivering prepackaged food portions to consumers who feel they are too busy to shop for themselves (e.g., www.blueapron.com) (Meter & Goldenberg, 2017). Thousands of small and large firms have entered the market (e.g., Amazon Fresh at www.amazon.com), while long-standing businesses have failed (Chasan & Burgdorfer, 2009; Bonato, 2016).

Four stages of the adaptive cycle

Scholars have found patterns in how systems change. Authors (Holling, 2001; Gunderson & Holling, 2002) identify four stages in the adaptive cycles of natural and human systems: exploitation, conservation, release, and reorganization. An adaptive cycle, Holling (2001) argues, 'embraces two opposites: growth and stability on the one hand, and change and variety on the other' (p. 392). In the first phase, new order is created out of chaotic conditions, sparking growth. During the second phase, the system's ability to efficiently control processes has the consequence of diminishing its capacity to adapt to change. Pressures for change mount, releasing latent energies. After this period of restructuring, resources begin to coalesce in new ways, creating a new cycle of growth. In all four phases, uncertainty abounds, either within the system itself or its external context.

Stroink and Nelson (2013) analyse five food system initiatives in Northern Ontario using this framework. Although they view the mainstream food system as positioned at the height of the conservation cycle, moving towards a rigidity trap, with local food initiatives as the release of resources creating change and innovation, they also caution that 'progression through the cycle is not necessarily sequential' (p. 7). They further note 'when a system will move in and through each phase cannot always be predicted, and a shift from one phase to the next is gradual and perhaps only identifiable after the fact' (p. 8). Finally, they add,

> [k]nowing where an organization lies on the adaptive cycle may allow us to understand the context of its challenges and to leverage interconnections for growth, but it does not allow us to predict with any accuracy what shifts or developments are likely to occur in the future, as these remain the emergent outcome of many complex interactions. (p. 26)

Moreover, as Holling (2001) posits, in a complex system adaptive cycles are nested across levels of scale in what he calls a 'panarchy' (p. 398). This suggests that any given system may exhibit qualities of any of the four adaptive cycle stages at different levels at any point. His practical concern is that 'actions that would be appropriate at one phase of the cycle might not be appropriate at other phases' (p. 402).

Order interacts with disorder

Adaptive systems are not totally chaotic. Multiple authors state that the generative and emergent properties of complex systems derive from the interactions between order with disorder (Montouri, 2008; Morin, 2008). These create leverage points. Holland (2006) notes that 'all CAS that have been studied carefully exhibit lever points – points where a simple intervention causes a lasting, directed effect' (p. 6). This is because the 'complexity of living systems of people and nature emerges not from a random association of a large number of interacting factors rather from a smaller number of controlling processes' (Holling, 2001, p. 391). It is just such leverage that offers hope to local food systems leaders amid massive global economies. Yet Holland (2006) also cautions, '[t]here is no theory that tells us where or how to look for CAS lever points' (p. 6). That means that subjective factors, such as imagination, pattern recognition, intuition, and educated insight, prove useful (Morin, 2008). As Holling (2001) states, '[w]e are always left with best judgments, not certainties' (p. 391).

'Wicked problems,' or 'problem situations'?

Adaptation presents food system leaders with ongoing issues that are unlikely to attain a clear resolution. Two planners, Horst Rittel and Melvin Webber (1973) are credited with coining a term for this: a 'wicked problem'. They point out that casting complicated situations as 'problems' grew out of scientific methods that attempted to divide complex issues into components. However, they caution, '[t]he classical paradigm of science and engineering – the paradigm that has underlain modern professionalism – is not applicable to the problems of open societal systems' (p. 160). They further explain that this paradigm has been 'dominated by the pervasive idea of efficiency' that was not appropriate for work in societal systems (p. 158; see also Australian Public Service Commission, 2007).

Rittel and Webber (1973) add that in a quest to identify all of the major agents in societal systems, or to understand in a mechanistic manner how these operated, analysts often overlook 'the most difficult discussion of all: "What *should* these systems do?"' (p. 157, original emphasis). The supposed ideal of a cybernetic planning system, including 'statistically monitoring those conditions of the publics and the systems that are judged to be germane [and] feeding back information to simulation and decision channels' (p. 159) is fundamentally impossible to create. They conclude, '[w]e all know that such a planning system is unattainable ... It is even questionable whether such a planning system is desirable' (p. 159).

Rittel and Webber (1973) further state that 'there is no definitive formulation of a wicked problem ... The information needed to understand the problem depends on one's idea for solving it ... The problem can[not] be defined until the solution has been found' (p. 161). Now, 45 years later, this could be put slightly differently, in part because they also state that 'social problems are never solved' (p. 160). Rather they are complex, ongoing issues (Flood, 1999; Chapters 1, 2, & 12, this volume).

Rittel and Webber (1973) add that no 'wicked' issue can even be *understood* without considering *all* of the potential solutions ahead of time. Our professional practice suggests that such complexity cannot be addressed until one has sufficiently constrained one's own view of the system by setting boundaries for the analysis as well (for example, by establishing geographical limits, knowing one's purpose, clarifying one's world view, or defining a clear focus for activity). Once these are defined, it is possible to identify specific patterns of emergence reflecting these choices.

Rittel and Webber (1973) further state that even the quest for an overarching method for addressing complexity might be illusory. Since there is 'no ultimate test of a solution to a wicked problem' (p. 164), there is no way of knowing all the repercussions of an action taken today. Nor are there 'criteria which enable one to prove that all solutions to a wicked problem have been identified and considered' (p. 164). Thus, in addressing the wicked problem of assuring food access for low-income residents while upholding farmer income, Nelson and Stroink (2014) conclude 'it was important to nurture connectivity that was not predefined by existing structure or imposed by outside leaders' (p. 10) in order to allow new approaches to emerge.

Each setting is unique

Rittel and Webber (1973) further caution that 'each wicked problem is essentially unique ... every situation is likely to be one-of-a-kind ... [therefore] one should not try to cure symptoms: and therefore one should try to settle the problem on as high a level as possible' (p. 165). Yet higher-level solutions cannot overlook local realities. As Morin (2008) states, complex thinking 'is always local, situated in a given time and place' (p. 97). Rittel and Webber (1973) add,

> [s]ocial science has simply been unable to uncover a social-welfare function that would suggest which decisions would contribute to a societally best state. Instead, we have had to rely upon the axioms of individualism that underlie economic and political theory, deducting, in effect, that the *larger-public* welfare derives from summation of individualistic choices. And yet, we know *this* is not necessarily so, as our current experience with air pollution has dramatized.
>
> (p. 168)

They conclude that

> in such fields of ill-defined problems and hence ill-definable solutions, the set of feasible plans of action relies on realistic judgment, the capacity to appraise 'exotic' ideas and on the amount of trust between planner and clientele that will lead to the conclusion, 'OK, let's try that'.
>
> (p. 164)

Flood (1999) goes on to state that

> [t]he idea of systemic thinking ... is to develop whole appreciation ... [which] may be achieved only through wide ranging discourse about the relevance of optional organisational structures *in each local context,* local in space and time ... People must be allowed to decide for themselves.
>
> (p. 72)

Flood (1999) adds, citing C.W. Churchman, 'systemic thinking (i.e. learning) begins when you see through the eyes of another' (p. 69).

If earlier scholars highlighted the uniqueness of each context, later work has added an emphasis on examining each situation through a diverse set of viewpoints, as well. Moreover, there are multiple valid approaches to working systemically. As Midgley (2007) points out, the very complexity of adaptive systems means that a complex variety of tools is useful, often in different phases of the same work. Even schools of thought or methodologies that became displaced by newer approaches may prove critically useful, and become refined, long after they have been discarded as 'obsolete'.

As a result, addressing persistent, complex food system issues in community contexts requires methodologies that tap into residents' immersed knowledge of place, both offering and creating opportunities to better understand recurring patterns they experience in their local context (Chapters 2, 3, 6, 7, 8, 9, & 10, this volume). Defining the complexity of food systems as a 'wicked problem' does not transcend the narrowness of defining 'problems' as if they were external conditions to be solved. Rather a CAS approach suggests that we are all – professionals are community members alike – immersed in complex settings. We need to embrace the complexity of systems, reach out to those with different viewpoints, learn from each other, and keep an eye out for changing conditions. In paying close attention to unique local conditions, we have found that often broader societal dynamics are illuminated as well.

Avoiding reductionism

Given the ongoing nature of complex social concerns, Aubin et al. (2013) clarify that '[t]he challenge facing assessment is thus to take full account of this systemic complexity without breaking it down into different parts, which would cause it to lose its interaction characteristics' (p. 200). Folke et al. (2005) add, '[a]daptive governance systems often self-organize as social networks with teams and actor groups that draw upon various knowledge systems and experiences for the development of a common understanding and policies' (p. 441). Similarly, Viljoen and Wiskerke (2012) place strong emphasis on network-building initiatives (see also Chapter 3, this volume). Kania and Kramer (2011, 2013) also place strong emphasis on self-organized networks. They conclude that no single 'backbone organisation' is essential for collective impacts; networks can effectively coordinate in decentralized manners (Meter, 2014). Folke et al. (2005, citing Ostrom, 2005) add, there is a 'need to view all policies as ongoing learning experiments that need to be monitored, evaluated, and adapted over time' (p. 447). As Morin (2008) puts it, '[d]on't forget that reality is changing, don't forget that something new can (and will) spring up' (p. 57).

Methodological frameworks

The approaches we have developed are essentially geared to assisting people who are inside a complex adaptive system to better understand, *from their diverse viewpoints and unique context* (Chapters 1, 2, 3, 6, 7, 8, 9, 10, & 11, this volume) how systems function and are adapting, sufficient to allow them to act without being paralysed by the complexity they experience. It is our aim to assist local practitioners to build their own capacity to self-organize within changing environments.

In our practice, we are invited in by a core of local leaders who have raised money to engage us. During our negotiations, we insist that local stakeholders define the issues to be addressed (Savin-Baden & Major, 2013; Hay, 2016).

Our approaches are asset-based (Kretzmann & McKnight, 1993; Goldman & Schmalz, 2005). We avoid doing 'needs assessments', since 'needs' discussions often do not move beyond what is lacking. We begin by learning what local leaders have already done, and asking what capacities the community already has, then identifying gaps and capacities they would like to strengthen.

We make one further distinction, viewing our approaches as *systemic*, rather than *systematic*, in nature (Flood, 2010). *Systematic* approaches strive for a comprehensive overview of a given system, often couched in hopes of modelling systems behaviour by considering all of the known elements of the system with quantitative rigor. As Rittel and Webber (1973) noted above, this is often impossible in a complex and adaptive environment. Rather, these methods strive to identify *systems dynamics*, including patterns of emergence, and specific attractors (Williams, 2005), as defined below.

We define *emergence* as unexpected patterns of complexity that result from simpler processes, but could not be predicted from the rules followed in the less complex process (Flood, 1999; Meter, 2007b; Stroink & Nelson, 2013). Rotmans and Loorbach (2009), citing De Haan (2006), distinguish three types of emergence: discovery, mechanistic, and reflective. The second is more structured than the first. In the latter type, 'observers are among the objects of the system, and have some reflective capacity, which enables them to observe the emergence they produce' (p. 3). Flood (1999) considers spontaneous self-organization to be a specific form of emergence. *Attractors* are patterns, clusters of energy, or resource flows that tend to create stability among disorder, and that may provide the backbone for lasting systems change (adapted from Eoyang, 2004; Meter, 2007b).

Participative research processes

Our approach to research in collaboration with community partners draws upon elements of Participatory Action Research (PAR), though it differs from more carefully constructed academic efforts. Cornwall and Jewkes (1995) identify four types of PAR: *contractual*, in which people are asked to participate in projects selected by researchers; *consultative*, in which people are consulted before interventions are made by outsiders; *collaborative*, referring to projects initiated by researchers, but carried out in collaboration with local people; and *collegiate*, in which researchers and residents work together as colleagues with different skills to offer. They add, '[w]hat is distinctive about participatory action research is not the methods, but the methodological contexts of their application' (p. 1667). They also focus considerable attention on the attitudes researchers bring with them, and their willingness to affirm the perspectives of local actors.

We operate almost entirely within the 'collegiate' realm. Although we are often handed considerable power by our clients in framing a research project, we do so to address concerns that the community has raised. As Cornwall and Jewkes continue, '[t]hrough a process of mutual learning and analysis, which takes part throughout research rather than at distinctive stages, people are brought into the research as owners of their own knowledge and empowered to take action' (p. 1670).

Yet our research process is often not collective in the ways that Baum et al. (2006) pursue. Often, rather than placing people in a room together in an effort to devise group analysis, we typically interview people privately, one-by-one. People typically offer more searching insights if they are speaking privately, reflecting on actions they have taken or observed in an exercise of praxis (Friere, 1972) (to understand the benefits of participatory research see Chapters 2, 3, 6, 8, 9, & 10, this volume).

Baum et al. (2006) note that '[t]he observer brings an impact on the phenomena being observed and brings to their inquiry a set of values that will exert influence on the study' (p. 854). We certainly make every attempt to enter each community discussions with limited preconceptions, mindful of Rocheleau's (1994) suggestion that

> [w]e have also found that there is no single best, fixed land use 'package' for any given region or group of people, but rather a vast array of principles and components that can be constantly recombined, tested, and modified to suit changing social, economic, and ecological conditions for individuals, households, communities, and nation-states. (p. 5)

Rocheleau further cautions (1994) that '[s]ome of our best data and insights are transmitted through stories, a professional oral tradition, and through the skills of our trades' (p. 9). This also applies to lay observers. Citing Hope and Timmel (1984), Rocheleau (1994) adds that '[f]armers, herders, and forest dwellers may benefit substantially from a combination of historical analysis, consideration of possible futures, and qualitative comparison of existing practice' (p. 13; see also Chapters 1 & 12, this volume).

Another way to ensure that multiple perspectives are given full consideration is to integrate mixed methods into the research process. Shorten and Smith (2017) point out that '[p]urposeful data integration enables researchers to seek a more panoramic view of their research landscape, viewing phenomena from different viewpoints and through diverse research lenses' (p. 74). In this way, groups of people consulting multiple data sources of data are more likely to embrace diverse sets of insights (Chapter 3, this volume).

Irwin (2008) reminds that quantitative and qualitative data are not distinct categories of data, they provide 'different *dimensions* of unitary problems… We need to keep under reflexive and critical scrutiny the categories we use to organize our thinking and order our data' (pp. 415–416, original emphasis). She continues, 'we do not properly understand the nature of people's values and subjective orientations if we do not understand the contexts in which they hold meaning' (p. 420).

Methodological tools

Time-series data

Time-series data are very valuable in identifying patterns over time (Rocheleau, 1994; Eoyang & Berkas, 1998; Aubin et al., 2013). These have turned out to be the key quantitative tool in our work, although surveys are also important for generating site-specific data. For work in the US, we draw heavily upon data from the US Bureau of Economic Analysis (BEA). These data are especially useful because they focus on income flows and do not take a position regarding production systems, programmes, or specific issues. Rather, they are intended to assist municipal leaders to understand local conditions. They provide annual county-level estimates of population, employment, personal income, and transfer payments, and are the only national source of annual local farm income data. Moreover, BEA data are reported for each county in the US for each year from 1969 to the present. They can readily be aggregated to provide income estimates for multi-county regions, including those that cross state boundaries.

Many similarly useful secondary data sets are available in the US, Canada (StatCan), the European Union, and elsewhere, but detailed local-level data can be difficult to locate in many countries. In the US, the Bureau of Labor Statistics Consumer Expenditure Survey, The Federal Census, and the Centers for Disease Control and Prevention health data have been highly useful in connecting farm income to demographic, food consumption, and health risk factors. Multiple data sets are also provided by the US Department of Agriculture (USDA), including the Census of Agriculture (which does include county-level acreage, production, sales, and other data in five-year increments); and state and national production and marketing data from the Economic Research Service (ERS), the Agricultural Marketing Service (AMS), and the Agricultural Resource Management Survey (ARMS). We also draw heavily upon GIS data from a number of public agencies for mapping land use, water features, topography, and transportation routes (Thilmany McFadden et al., 2016).

Limits to data and interpretation

BEA data do have limitations. They are not raw counts, but rather estimates modelled by professionals with access to considerable computing power, drawing data from a variety of sources. This means county-level data may reflect estimates drawn from state, regional, or national level averages. However, BEA correlates these to primary business filings so we have found that local data reflect local conditions fairly well. As with any time-series, often the patterns and trends depicted are more meaningful than the specific data points.

Interpretation should also be performed with care. There have been many structural changes to the farm economy since 1969. Far fewer farmers are in business now, and the level of technology is far more intense. BEA time-series data portray the farm sector as a whole. While this is important to know, fine-grained calculations require additional data sets in order to better understand how conditions compare across time.

A further limitation is that most national secondary data sets show primarily the workings of large-scale farms and the commodity industry, not trends in community food production (Goldenberg & Meter, 2019). These broader trends do strongly affect those producing food for nearby consumers, influencing the cost of land, availability of labour and other farm inputs, opportunity costs, and a host of other concerns, but once again it is important to interpret what *is* shown carefully. Until comprehensive and accurate data sets (or suitable proxies) are compiled covering community food production, these broader sets will be the best available. Importantly, time-series data often show that the commodity economy is not as robust as people imagine (see Figure 4.1). This often fosters a *destructuring* (Nelson & Stroink, 2013) of the assumptions that community members make about the mainstream food system, as the next section discusses.

A century of US farming

One example will demonstrate what time-series data can portray. In Figure 4.1, more than 100 years of US farm income and production expense data are shown. In this case, to consider the national context, we refer to USDA Economic Research Service data (taken from ERS Farm Income and Wealth Statistics), which show longer trends than BEA data. Included are (a) cash receipts; (b) production expenses; and (c) net cash income (cash receipts less production expenses – a measure of the profitability of producing crops and livestock) for all farmers in the US, from 1910 to 2018. These data have been adjusted for inflation using the US Federal Reserve Bank's consumer price index (CPI) (US Federal Reserve Bank of Minneapolis, n.d.) and are presented in 2018 US dollars.

The unadjusted data show long-term increases in both cash receipts and expenses. Yet richer insights emerge after adjusting for inflation. Visible in Figure 4.1 are several periods of prosperity for the US farm sector: (a) the 'Golden Age' of 1910–1914 when the US dominated global commodity markets and living costs were low, still viewed as the standard for farm prosperity, but followed by sales spurred by the First World War; (b) the period during and immediately after the Second World War when wartime and recovery-era sales and new technology combined to boost farm income; (c) the OPEC oil crisis of 1973–1974, when the US sold massive

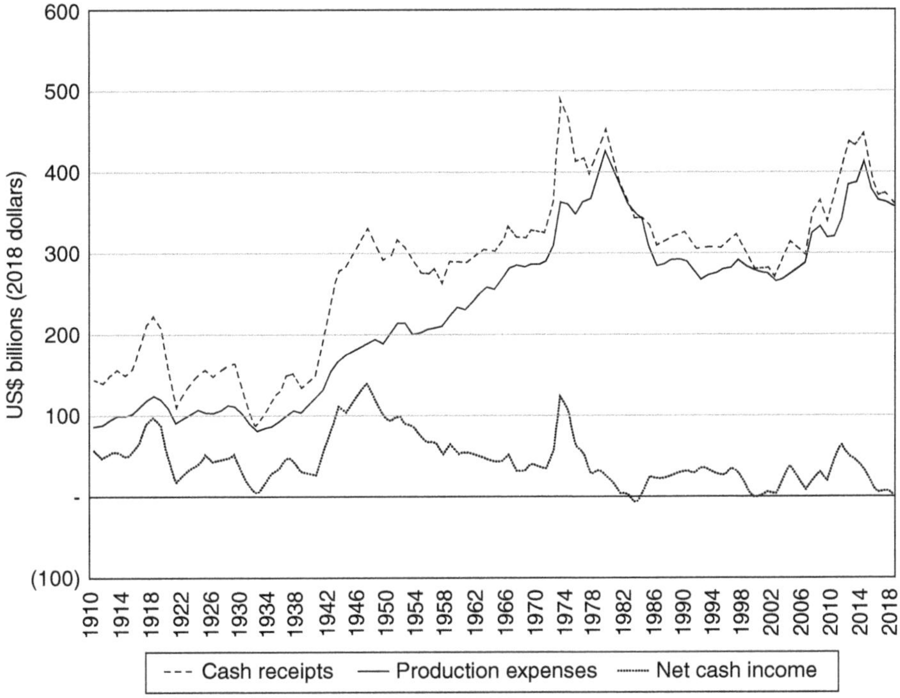

Figure 4.1 Adjusted Net Cash Income for US Farms, 1910–2018.

Source: US Economic Research Service Farm Income Balance Sheet data. Public domain. Adjusted for inflation using US Federal Reserve Consumer Price Index.

shipments of grain to the USSR in order to retrieve its currency; and (d) the global housing debt crisis of 2008–2011 when speculators' bidding up the price of grains and rising ethanol use conspired to create a commodity price bubble.

What this chart shows, then, is that the only truly prosperous periods for US agriculture were due to either long-lost global market power (a) or to external shocks (a–d). Overall, net cash income trended downward, reaching lower levels in 1983, 2000, and 2018 than during the Great Depression (1932). Given the considerable importance of federal subsidies, tax incentives, and infrastructure investments to encouraging long-distance commodity trade, this is a stark reality. Moreover, each bubble led to further decline – not to lasting prosperity for the farm sector. More fundamentally, farmer investments in substantial new productivity (ERS, Agricultural Productivity in the US, series) yielded little return.

What this chart does not show is that off-farm income has increased dramatically over the past century (ERS, Farm Household Income and Characteristics, series). So, while farm families may have acceptable incomes (Gardner, 2002), especially if they purchased land a generation or more ago, they increasingly earn this off the farm – not by actually producing commodities. In fact, ERS now calculates that US farmers lose more than US$60 per acre planting corn (ERS, Corn Production Costs and Returns, series). Each of these trends shows adaptive cycles within the prevailing commodity economy, a context that creates considerable uncertainty despite federal efforts to stabilize.

Since key insights like these can be drawn from readily available public data sets, we often wonder why so many analysts overlook them. The patterns that local practitioners recognize in these secondary data sets often serve as a starting point for further critical reflection on the part of both researchers and community members.

Interviews with wise practitioners

This leads to the most essential qualitative tool: interviews with 'wise practitioners'. These are people well immersed in their context, who have gained special insight, or who are especially articulate in taking a broader view of the issues at hand (Meter, 2007b). They intimately know the issues that people face on the ground, and yet they also take a position that is broader than their personal self-interest. As Flood (1999) states, they have 'learned inside feedback structures' (pp. 68–69), self-reflective and also responsive to those around them.

The most informed have also engaged in thoughtful reflection about their own mental frameworks (Flood, 1999; Morin, 2008). In short, they are able to 'see the world through the eyes of another' (Flood, 1999, p. 63). Often their own efforts to change prevailing power dynamics offer them rich insight into systems dynamics, including how a system pushes back against change (Maretski & Tuckermanty, 2007; Nelson & Stroink, 2014). Often the most informed practitioners are generalists, rather than specialists (Rocheleau, 1994; Flood, 1999; Meter, 2007b; Morin, 2008; Stroink & Nelson, 2013; Chapter 11, this volume).

Farmers, low-income residents, and other marginalized food system stakeholders are deeply cognizant of complexity. They recognize they hold limited power in making conditions better, and have often experienced potent resistance from prevailing systems. This pushback illuminates critical system dynamics (Meter, 2007b). Speaking primarily of culturally embedded knowledge in ecology, but also with clear application to social systems, Folke et al. (2005) argue that

> [k]nowledge acquisition of complex adaptive systems is an ongoing, dynamic learning process ... It comes as no surprise that knowledge of ecosystem dynamics and associated management practices exists among people of communities that, on a daily basis and over long periods of time, interact for their benefit and livelihood with ecosystems.
>
> (pp. 445–446)

One central challenge is to identify the most informed wise practitioners to interview. In our practice, we ask a group of local food system leaders to select the best people to interview based on a collective sense of which people have: (a) done the most to build the food system that embodies a more sustainable future; or (b) hold the most penetrating insights into how community foods efforts have played out in the past.

If the members of the group that make such a selection trust each other well, and keep an open mind to the broader purpose of the study, essential system dynamics and levers for change may easily be revealed from a relatively small number of interviews (about 15), unless the context is especially conflicted. In particular, primary producers (farmers, fishers, processors, etc.) often hold the most informed insights about broader aspects of the food system since their point of view is both pragmatic and immersed, and also informed by cultural roots. Independent business owners or managers may also hold a solid sense of local tradition that balances their attention to markets. Similarly, low-income residents or others who are marginalized often know more about systems dynamics than those who are in power, because they have confronted social systems and witnessed the resistance (Friere, 1972; Meter, 2007b; Morin, 2008).

Additional interviews may also be essential to fully account for diverse viewpoints, or for credibility's sake. We often discover that key practitioners have been overlooked by the local leadership team. Ideally, no key player, industry sector, nor any key philosophical position taken within the community should be excluded. Several clients have also pointed out that unless local people are directly quoted in a food system assessment, civic leaders will have limited interest in the results. This may reflect a cynicism about modelled, or academic, analysis, but it also expresses political pragmatism as civic leaders respond to constituents.

Essentially, this interview process asks those who have extensive experience in the food system's complex adaptations – its emergent patterns and its attractors – to identify the 'levers' that would best move the system. Since these systems levers are expressed in the words of local leaders, the public often understands them readily. Yet practitioners' views are not always accurate, so the quantitative data compiled and the professional experience of the analyst both play key roles in ultimately defining which levers would move the system to fulfil the local initiative's purposes at the current stages of the adaptive cycles. Those community members who implement the systemic strategies selected will further refine this analysis over time, and may come to reject the consultants' views. As systemic work in a complex adaptive setting, the main importance is that professionals and community members learn together, and that local stakeholders define and select from the options of strategic choices. As stated above, it is the trust and openness developed in this relationship, more than the data provided or the models applied, that will make the critical difference.

Developing strategic direction

Avoiding needs assessments, we often use a SWOT (Strengths, Weaknesses, Opportunities, Threats) framework. We always begin with asking about 'strengths'. We commonly will list these attributes of the local food system anonymously to separate personalities from concepts. We look for clues that will suggest a vision for systemic change, rather than pointing to a specific programme (Morin, 2008). Often these SWOT depictions also suggest systems levers, identify emergent properties and attractors, or help us understand how comprehensive local perspectives on the prevailing food system may currently be.

Setting up a food system assessment

There are two key decisions that must be made to begin an assessment of a complex adaptive food system. In our practice, we ask local partners to make these determinations. The first decision is to make the purpose of, and audience for, the study explicit. The second is to define a specific geographic scope. Often these two decisions must be made iteratively. As noted above, these two decisions both frame and limit the analysis of the system, clarifying which elements of the system will take priority (for other approaches, refer to Chapters 2, 8, & 9, this volume).

Make the purpose and audience explicit

In our initial agreement, we specify that one of the purposes for the assessment is to build the capacity of local residents to strengthen and manage their own initiatives. Further, we ask the client what they hope to accomplish with our final report. Do they seek a document that can sway commissioners in one particular county? Is the purpose to expand regional food trade within a 32-county region? Is the purpose to explore the feasibility of a specific business opportunity? Each potential purpose suggests its own approaches, inquiry, and methods.

One important distinction we are increasingly making is to suggest that a project's purpose be set as building 'community-based food systems' (CBFS), not just 'local' or 'regional' systems. For years, the concept of 'local food' has been diluted by practitioners who define it to mean whatever they can conveniently deliver to their customers in a way that will not invite too much scrutiny. We have come across definitions of 'local' that range from 'within my township' to 'anywhere we can drive within 24 hours'. Such mileage limits are easy to co-opt, while building social capital is what provides a competitive advantage to community foods efforts (Goldenberg & Meter, 2019).

We define CBFS as *systems of exchange that strive to bring farmers and consumers into affinity with each other* (Heffernan, 1999; Maretski &

Tuckermanty, 2007; Stevenson et al., 2011; Wiskerke & Viljoen, 2012), for the purposes of building health, wealth, connection, and capacity (Meter, 2003, 2007b).[1] This definition views culture as a meta-purpose integrating a community's attention to the other four purposes, and serving as the central vehicle through which people agree to share certain rituals – including food – that connect them to place, and regenerate social memory (Folke et al., 2005). Folke et al. (2005) define 'social memory' as 'the arena in which captured experience with change and successful adaptations, embedded in a deeper level of values, is actualized through community debate and decision-making processes into appropriate strategies for dealing with ongoing change' (p. 453). At a more abstract level, Morin (1992) calls for 'face-to-face contact' between the observer and the system to be observed, and points out that culture 'provides the paradigms which permit and require' (p. 10) acts of distinguishing.

Indeed, using the very term 'community-based food system' points to the complexity of the discussion because community is complicated. It appears to be easier to simply count food miles or define a rigid geography. Yet a complexity focus also heightens opportunity for success, in our experience. It assumes that an effectively self-organizing community will pay attention to water, soil, and air quality for the sake of human health. While this is an implicit promise of democracy (Putnam, 1993; Lyson, 2007), such care is not always in force. Moreover, the weight of prevailing economic infrastructure in the US operates to frustrate the possibility of community-based enterprise (Meter, 1990; Heffernan, 1999), so this focus invites learning about negative feedback loops.

Define geographic boundaries

The second key question is to define the geography that the assessment will cover. This is inherently a local decision in the case of a local food initiative (Lynch et al., 2015; see also Chapter 9, this volume). It is also an arbitrary one. The *only* geographic boundaries that are not arbitrary are island or watershed boundaries, because these are determined by ocean frontage or the peak elevation of the surrounding landscape. Many work with a seemingly parallel concept of a 'foodshed', defined by Kloppenberg et al. (1996) as 'streams of foodstuffs running into a particular locality, their flow mediated by the features of both natural and social geography' (p. 12). Yet boundary decisions are complex, because the distance food can conveniently travel is determined as much by the presence of freeways, bridges over waterways, warehouses, and airports as by topography. Thus, the term 'foodshed' is a useful metaphor, if it is not taken too strictly.

For simplicity in regional identity, data gathering, or writing policy it may be useful to work within county or state boundaries, yet food systems routinely transcend such distinctions. Moreover, many regions contain big discontinuities in arable land. The key is that the boundaries should reflect the scope of the current or proposed food system, rather than bureaucratic or organizational lines.

In general, we find that the 'best' geography is the one in which the client has a mutually respectful dialogue established with key food systems leaders. If the geography is smaller than a US county, this makes secondary data sets less useful, but primary data can still be compiled. If the region is defined as a larger area to enlarge the number of potential consumers, the questions become: how far does your trust extend? If you called a meeting, who would actually show up? Who would think it is too far to drive?

Once these two elements are clear, and the boundaries of the systemic view have adequately been set, the focus has been clarified enough that it is possible to identify emergent properties, attractors, and systems levers. After initial consideration, we propose a set of activities to the client, and they decide whether this is a good match for their needs.

Using the community capitals framework

We believe this approach is quite consistent with the work of Cornelia and Jan Flora, who in particular have developed and applied a 'community capitals framework' that lists human connectivity as one of seven forms of capital including natural, human, cultural, financial, built, and political (Flora & Flora, 2004; Chapter 3, this volume). Economic flows integrate across all of these capitals (Morin, 2008), as does social capital. Over time we have placed stronger focus on the need for strong social capital to create a resilient community economy (Goldenberg & Meter, 2019). Holling (2001) adds that 'connectedness ... determines the degree to which a system can control its own destiny' (p. 394). Folke et al. (2005) continue, '[s]ocial sources of resilience, such as social capital (including trust and social networks) and social memory (including experience for dealing with change) are essential for the capacity of social-ecological systems to adapt and shape change' (p. 444).

Connecting to sustainable food systems work

This CAS approach overlaps considerably with other sustainable food systems work. The challenge in each community setting is for self-organizing activity to align itself with its environmental context to accomplish sustainability goals. This suggests that data providers should provide solid measures to inform strategic discussions that must inherently address very unique local conditions.

Global and national data sets covering the panoply of sustainability concerns are essential for informing these local systemic initiatives. On the other hand, if global data initiatives are designed as an effort to limit local decision-making, or to unify efforts across geographies, they may interfere with community-based efforts to move systems levers that are profoundly rooted in place (Morin, 2008).

Channelling results into a workplan

As outsiders seeking to define processes that are inherently internal to a specific community, and most visible to those who live inside the context, and as researchers who hope to build the capacity of our partners, our work converting research insights into practical workplans can be a precarious process. The basic insight to keep in mind is that there is no magic single answer. Nor can success in a complex adaptive food system be understood by referring only to a single measurement, because doing so distorts the complex interactions within the system and leads to harmful consequences (Midgley, 2007). Moreover, implementation in an adaptive system involves trade-offs between competing forces. Holding a complex set of purposes may help encompass these potential tensions. Keeping a complexity paradigm at the forefront of the work, it is possible to address systems levers and emergent change, while placing a lower priority on other critical concerns (see Chapters 2, 3, 8, & 9, this volume).

Typically, the proposed strategy is itself complex. We often compile a list of perhaps 5–8 proposed action steps, knowing that each one suggests pushing one or more 'levers' to create systemic change. As just one example, we often suggest that local leaders build greater trust and form stronger collaborations, assuming that stronger social capital will foster greater resilience (Meter, 2012; Meter, 2016; Meter & Goldenberg, 2018; Goldenberg & Meter, 2019).

Discrete progress in moving one of these systems levers may result in larger impacts over time, but may simply evoke resistance. If community members recognize their own wisdom in the systems levers that have been identified, however, they are in a solid position to ride through these uncertainties. Still, offering complex choices may feel debilitating to some community initiatives that seek simple solutions.

Measuring success using linked indicators

Having defined systems levers, and making a strategic assessment of which ones to push at the present time, measuring success is cast in different terms. To identify levers is essentially to begin to define a *systemic theory of change*. That is to say, a theory of change that: (a) takes into account the adaptive dynamics of prevailing food systems, and (b) measures whether systems levers have actually been moved, and if so, (c) whether this has shifted system dynamics. When issues are highly linked – as they always are in a CAS – systemic indicators should also express these linkages. For example, building social connectivity can become a foundation for economic and health improvement (Lynch et al., 2015; Goldenberg & Meter, 2019).[2]

One of the key systems dynamics that has been identified in regional farm and food economy studies is that the prevailing food system is structured to extract wealth from urban low-income areas and rural communities in the US (Goldschmidt, 1978; Meter, 1990; Meter, 2007b) as well as globally (Stevenson et al., 2011). Finding local systems levers that will effectively reverse this extraction is an exceptional challenge (see Chapters 5 & 11, this volume). Yet operating a community foods initiative without addressing these structures strategically runs the risk of deepening the extraction (Meter, 2007b).

Implementation of the plans

Tragically, implementation in the US is frequently stalled by a combination of factors: years of partisan stalemates have reduced the budgets available at the local level for communities to act on their own behalf. Federal aid to municipal governments has been cut, and political discourse even at the local level has become polarized. Many citizens have opted to reject public investment of any kind.

We try to define the most comprehensive strategy possible, based on what seems right to do to move a community towards its community foods purposes. Typically, this means that leaders will make use of the data and the recommendations for years to come, adapting their strategies as conditions change.

Impacts of CAS assessments

Two studies stand out as exemplars of the utility of taking a complex adaptive view. First, when the Indiana State Department of Health commissioned an assessment of the Indiana food system so the agency could better learn how to effectively intervene to promote better health, our interviews documented that state policy was focused on exporting food globally, rather than meeting the needs of state residents (Meter, 2012). Meanwhile, state residents who were creating community alternatives were marginalized. The report presented these community-based food initiatives *as an emergent quality of the prevailing food system*, rather than as a separate cluster of activity. This, of course, is consistent with Holling's (2001) 'release' cycle. This framing made it very difficult for political leaders to deny their presence without rebuffing their own constituents. Moreover, quantitative data showed that while Indiana farms exported more than US$8 billion of wholesale commodities, the state's consumers purchased an estimated US$14.5 billion of food items sourced outside the state each year. This comparison deepened the discourse about food in Indiana. That is to say, our research identified two central systems levers: challenging the notion that Indiana 'feeds the world', and questioning whether the state benefited economically by attempting to do so.

Raising these questions deepened the discussion of food and agriculture in Indiana. Rural legislators seized on the amount of food the state imported each year (Indiana Rural Caucus, 2012). The head lobbyist for the Indiana Farm Bureau called this report his 'bible' (White, 2014). The state's land-grant university moved to hire a local foods coordinator.

That report's recommendation to invest in strengthening community foods networks was adopted in a second study by economic developers in an 11-county region of Northeast Indiana (Meter, 2016). In this case, one systems lever was ensuring that food leaders adopted a single definition of 'local food network'. Ultimately, our analysis brought economic developers into stronger engagement with issues of poverty and farm business development (Goldenberg & Meter, 2019). Thus, they began to embrace emergent farm business networks, and deepened their view of network-building as a path forward. Thus, the state Health Department initiative influenced efforts to build social capital in two settings (see Chapter 11, this volume).

In turn, advances made in Indiana impacted other states. Northeast Indiana's embrace of network-building helped inspire the Maricopa County (Arizona) Food Systems Council to commission a study to show how community foods activity was building social and commercial networks in the Phoenix metro region. Our report (Meter & Goldenberg, 2018) documented the growth of community networks, and also noted how these contributed to local economic multipliers. Our social network maps showed that food buyers played a significant role in building social capital in the region, often more centrally than farmer organizations or non-profits. This network-building approach has been embraced by both the local food systems council and by civic leaders.

Limitations of this approach

The approach outlined here does not satisfy those who want easy answers or a simple formula they can follow (for more about complexity approaches, see Chapters 2, 3, 5, 6, 8, 9, 10, & 11, this volume). Those who seek such a linear answer operate in a different reductionist sphere, perhaps because they are not ready to embrace the complexity of the communities where they work. Our experience is that when people opt for simpler, more linear approaches, greater dysfunction is often created.

Our focus on economics could also appear too narrow. Yet for better or worse, economics is the language the US typically uses in making key policy decisions, and *is* how societies value the trade-offs between competing environmental and social goals. Like any other discipline, of course, economics is itself a complex adaptive system, yet its role is also somewhat unique. As Morin (2008) writes, 'The economic dimension contains the other dimensions and there is no reality we can comprehend with a single dimension' (p. 46). Our approach endeavours to create an economic language that helps explain a broader reality and embraces positive change rather than deepening exploitation.

Adaptive approaches posit questions to be addressed through ongoing reflection, but do not offer definitive solutions. While identifying systemic levers and emergent properties may help simplify assessment, this does not mean that food systems become simple. As Morin (1992) sums it up: '[c]omplexity cannot be simplified' (p. 11). But using a complexity paradigm may well show a more concise path for local leaders to address convoluted challenges for decades to come.

Notes

1 The careful reader will note that this formulation addresses Rittel and Webber's call to define a purpose for a system. Moreover, as a *system* of purposes, it is itself a complex adaptive system attuned to consideration of CAS dynamics. The concept of Community-Based Food Systems certainly also reflects Lyson's pioneering work (Lyson, 2004).

2 Some approaches for defining and using linked indicators are covered in Meter (1998, 2007a, p. 183).

References

Allen, T. & Prosperi, P. (2016) Modeling sustainable food systems. *Environmental Management*. 57, pp. 956–975.

Allen, T., Prosperi, P., Cogill, B., Padilla, M., & Peri, I. (2018) A Delphi approach to develop sustainable food system metrics. *Social Indicators Research*. 36(1), pp. 3–19. Available at: https:doi.org/10.1007/s11205-018-1865-8.

Aubin, J., Donnars, C., Supkova, M., & Dorin, B. (2013) A critical panorama of methods used to assess food sustainability. In: Esnouf, C., Russel, M., & Biracs, N. (Eds.) *Food system sustainability: Insights from duALIne*. Cambridge: Cambridge University Press.

Australian Public Service Commission (2007) *Tackling wicked problems: A public policy perspective*. Report available at: https://www.apsc.gov.au/tackling-wicked-problems-public-policy-perspective.

Baum, F., MacDougall, C., & Smith, D. (2006) Glossary: Participatory action research. *Journal of Epidemiology and Community Health*. 60, pp. 854–857.

Berkes, F., Folding, J., & Folke, C. (2000) Rediscovery of traditional ecological knowledge as adaptive management. *Ecological Applications*. 10, pp. 1251–1262.

Blay-Palmer, A., Santini, G., Dubbeling, M., Renting, H., Taguchi, M., & Giordana, T. (2018) Validating the city region food system approach: Enacting inclusive, transformational city region food systems. *Sustainability*. 10, p. 1680.

Bonato, G. (2016) New titans on the block: ABCDs lose top Brazil grains spot to Asian rivals. *Reuters*, 23 March.

Chasan, E. & Burdofer, B. (2009) Pilgrims pride exits bankruptcy under JBS deal. *Reuters*, 28 December.

Collona, P., Fournier, S., & Touzard, J. (2013) Food systems. In: Esnouf, C., Russel, M., & Biracs, N. (Eds.) *Food system sustainability: Insights from duALIne*. Cambridge, Cambridge University Press.

Cornwall, A. & Jewkes, R. (1995) What is participatory research? *Social Science and Medicine*. 41(12), pp. 1667–1676.

Crossroads Resource Center (n.d.) *About us*. Available at: www.crcworks.org.

De Haan, J. (2006) How emergence arises. *Ecological Complexity.* 3(4), pp. 293–301.

Eoyang, G.H. (2004) *Complex adaptive systems (CAS).* Battle Creek, MI, W.K. Kellogg Foundation.

Eoyang, G.H. & Berkas, T.H. (1998) *Evaluation in a complex adaptive system.* Circle Pines, MN, Human Systems Dynamics Institute, p. 2010. Available at: http://www.academia.edu/830274/Evaluation_in_a_Complex_Adaptive_System_April_30_1998.

Fabricius, C. & Koch, E. (Eds.) (2004) *Rights, resources and rural development: Community-based natural resource management in southern Africa.* London, Earthscan.

Feenstra, G. (1997) Local food systems and sustainable communities. *American Journal of Alternative Agriculture.* 12, p. 1.

Flood, R.L. (2010) The relationship of 'systems thinking' to action research. *Systemic Practice and Action Research.* 23(4), pp. 269–284.

Flood, R.L. (1999) *Rethinking the fifth discipline: Learning within the unknowable.* London, Routledge.

Flora, C.B. & Flora, J.L. (2004) *Rural communities: Legacy and change* (2nd ed.) Boulder, CO, Westview Press.

Folke, C., Colding, J., & Berkes, F. (2003) Synthesis: Building resilience and adaptive capacity in social-ecological systems. In: Berkes, F., Colding, J., & Folke, C. (Eds.) *Navigating Social-Ecological Systems: Building Resilience for Complexity and Change.* Cambridge, Cambridge University Press, (pp. 352–387).

Folke, C., Hahn, T., Olsson, P., & Norberg, J. (2005) Adaptive governance of social-ecological systems. *Annual Review of Environmental Resources.* 30, pp. 441–473.

Friere, P. (1972) *Pedagogy of the oppressed.* Harmondsworth, Penguin.

Gardner, B. (2002) *American agriculture in the twentieth century: How it flourished and what it cost.* Cambridge, MA, Harvard University Press.

Goldenberg, M.P. & Meter, K. (2019) Building multipliers rather than measuring them: community-minded ways to develop economic impacts. *Journal of Agriculture, Food Systems, and Community Development.* 8(3), pp. 153–164.

Goldman, K.D. & Schmalz, K.J. (2005) 'Accentuate the positive!' Using an asset-based mapping tool as part of a community health needs assessment. *Health Promotion Practice.* 6(2), pp. 125–128.

Goldschmidt, W. (1978) *As you sow: Three studies in the social consequences of agribusiness.* Montclair NJ, Allanheld, Osman.

Gunderson, L.H. & Holling, C.S. (2002) *Panarchy: Understanding transformations in human and natural systems.* Washington, DC, Synopsis published by Island Press.

Hay, I. (2016) *Qualitative research methods in human geography.* Oxford, Oxford University Press.

Heffernan, W. (1999) *Consolidation in the food and agriculture system.* University of Missouri-Columbia, Report to National Farmers Union, 5 February.

Holland, J.H. (1992) Complex adaptive systems. *Daedalus.* 121(1), pp. 17–30.

Holland, J.H. (2006) Studying complex adaptive systems. *Journal of Systems Science and Complexity.* 19, pp. 1–8.

Holling, C.S. (2001) Understanding the complexity of economic, ecological, and social systems. *Ecosystems.* 4, pp. 390–405.

Hope, A. & Timmel, S. (1984) *Training for transformation.* Volumes 1–3. Harare, Mambo Press.

Indiana Rural Legislators' Caucus (2012) *Meeting at State Capitol in Indianapolis,* 18 January.

Irwin, S. (2008) Data analysis and interpretation: Emergent issues in linking qualitative and quantitative evidence. In: Hesse-Bieber, S.N. & Leavy, P. (Eds.) *Handbook of emergent methods*. New York, Guilford Press.

Kania, J. & Kramer, M. (2011) Collective impact. *Stanford Social Innovation Review*. Winter. Available at: http://www.ssireview.org/articles/entry/collective_impact.

Kania, J. & Kramer, M. (2013) *Embracing emergence: How collective impact addresses complexity* (pp. 36–41). Available at: http://www.ssireview.org/pdf/Embracing_Emergence_PDF.pdf.

Kloppenberg, J., Hendrickson, J., & Stevenson, G.W. (1996) Coming into the foodshed. *Agriculture and Human Values*. 13(3), pp. 33–42.

Koopmans, M. (2017) Perspectives on complexity, its definition and applications in the field. *Complicity: An International Journal of Complexity and Education*. 14(1), pp. 16–35.

Kretzmann, J.P. & McKnight, J.L. (1993) *Building communities from the inside out: A path toward finding and mobilizing a community's assets*. Chicago, IL: ACTA Publications. Available at: https://www.amazon.com/Building-Communities-Inside-Out-Mobilizing/dp/087946108X.

Lynch, J., Meter, K., Robles-Schrader, G., Goldenberg, M.P., Bassler, E., Chusid, S., & Austin, C.J. (2015) *Exploring economic and health impacts of local food procurement*. Minneapolis, MN, Crossroads Resource Center; Chicago, IL, Illinois Public Health Institute. Available at: www.crcworks.org/EHimpacts.pdf.

Lyson, T. (2004) *Civic agriculture: Reconnecting farm, food, and community*. Medford, MA, Tufts University Press.

Lyson, T. (2007) Civic agriculture. In: Hinrichs, C.C. & Lyson, T.A. *Remaking the American food system*. Lincoln, NE, University of Nebraska Press.

Maretski, A.N. & Tuckermanty, E. (2007) Community food projects and food system sustainability. In: Hinrichs, C.C. & Lyson, T.A. *Remaking the American food system*. Lincoln, NE, University of Nebraska Press.

McIntosh, R.J. (2000) Social memory in Mande. In: McIntosh, R.J., Tainter, J.A., & McIntosh, S.K. (Eds.) *The way the wind blows: Climate, history and human action*. New York, Columbia University Press, pp. 141–180.

Meter, K. (1990) *Money with roots*. Minneapolis, MN, Crossroads Resource Center. Available at: www.crcworks.org/roots.pdf.

Meter, K. (1998) *Neighborhood sustainability indicators guidebook*. Minneapolis, MN, Crossroads Resource Center. Available at: www.crcworks.org/guide.pdf.

Meter, K. (2003) *Food with the farmer's face on it: Emerging community-based food systems*. Battle Creek, MI, W.K. Kellogg Foundation. Available at: www.crcworks.org/fface.pdf.

Meter, K. (2007a) Linked indicators of sustainability build bridges of trust. In: Maida, C.A. (Ed.) *Sustainability and communities of place*. Oxford, Berghahn Books, pp. 183–200.

Meter, K. (2007b) Evaluating farm and food systems in the U.S. In: Williams, B. & Imam, I. (Ed.) *Systems concepts for evaluation*. Point Reyes, EdgePress of Inverness, pp. 141–160.

Meter, K. (2010) Letting food systems emerge: How do we measure progress in an emergent system? *Journal of Agriculture, Food Systems, and Community Development*. 1(1), pp. 23–26.

Meter, K. (2012) *Hoosier farmer? Emergent food systems in Indiana*. Minneapolis, MN, Crossroads Resource Center. Available at: http://www.crcworks.org/infood.pdf.

Meter, K. (2014) Co-ops and collective impact. *Journal of Agriculture, Food Systems, and Community Development*. 4(3), pp: 11–14. Available at: http://dx.doi.org/10.5304/jafscd.2014.043.014.

Meter, K. (2016) *Northeast Indiana Local Food Network – Phase 1 Report*. Minneapolis, MN: Chicago IL, Manheim Solutions. Available at: http://www.crcworks.org/innetworks16.pdf.

Meter, K. & Goldenberg, M.P. (2017) *Hawai'i food for all*. Minneapolis, MN, Crossroads Resource Center. Available at: http://www.crcworks.org/hifood.pdf.

Meter, K. & Goldenberg, M.P. (2018) *Building community networks through community foods*. Minneapolis, MN, Crossroads Resource Center. Available at: http://www.crcworks.org/azmaricopa18.pdf.

Midgley, G. (2007) Systems thinking in evaluation. In: Williams, B. & Imam, I. (Eds.) *Systems concepts in evaluation*. Point Reyes, EdgePress of Inverness, pp. 11–34.

Montouri, A. (2008) Forward: Edgar Morin's path of complexity. In: Morin, E. (Ed.) *On complexity*. Cresskill, NJ, Hampton Press.

Morin, E. (1992) From the concept of system to the paradigm of complexity. *Journal of Social and Evolutionary Systems*. 15(4), pp. 371–384.

Morin, E. (2008) *On complexity*. Cresskill, NJ, Hampton Press.

Nelson, C.H. & Stroink, M.L. (2014) Accessibility and viability: A complex adaptive systems approach to a wicked problem for the local food movement. *Journal of Agriculture, Food Systems, and Community Development*. 4(4), pp. 191–206.

Prosperi, P., Allen, T., Cogill, B., Padilla, M., & Peri, I. (2016) Towards metrics of sustainable food systems: A review of the resilience and vulnerability literature. *Environment Systems & Decisions*. 36, pp. 3–19.

Prosperi, P., Moragues-Faus, A., Sonnino, R., & Devereux, C. (2015) *Measuring progress toward sustainable food cities: Sustainability and food security indicators*. Cardiff, Cardiff University School of Planning and Geography. Available at: http://sustainablefoodcities.org/getstarted/developingindicators.

Putnam, R.D. (1993) The prosperous community: Social capital and public life. *The American Prospect*. 13(Spring), pp. 35–42. Available at: http://www.prospect.org/cs/articles?article=the_prosperous_community.

Rittel, H.W.J., & Webber, M.M. (1973) Dilemmas in a general theory of planning. *Policy Sciences*. 4(2), pp. 155–169.

Rocheleau, D.E. (1994) Participatory research and the race to save the planet: Questions, critique, and lessons from the field. *Agriculture and Human Values*. Spring–Summer, pp. 4–25.

Rotmans, J. & Loorbach, D. (2009) Complexity and transition management. *Journal of Industrial Ecology*. 13(2), pp. 1–13.

Savin-Baden, M. & Major, C.H. (2013) *Qualitative research: The essential guide to theory and practice*. London, Routledge.

Shorten, A. & Smith, J. (2017) Mixed methods research: Expanding the evidence base. *Evidence Based Nursing*. 20(3), pp. 74–75.

Stevenson, G.W., Clancy, K., King, R., Lev, L., Ostrom, M., & Smith, S. (2011) Midscale food value chains: An introduction. *Journal of Agriculture, Food Systems, and Community Development*. 1(4), pp. 27–34.

Stevenson, G.W., Ruhf, K., Lezberg, S., & Clancy, K. (2007) Warrior, builder, and weaver work. In: Hinrichs, C.C. & Lyson, T.A. (Eds.) *Remaking the American food system*. Lincoln, NE, University of Nebraska Press.

Stroink, M.L. & Nelson, C.H. (2013) Complexity and food hubs: Five case studies from northern Ontario. *Local Environment.* 18(5), pp. 620–635.

Swanson, D. & Bhadwal, S. (2009) *Creating adaptive policies.* International Development Research Centre. New Delhi, Sage.

Thilmany McFadden, D., Conner, D., Deller, S., Hughes, D., Meter, K., Morales, A., Schmit, T., Swenson, D., Bauman, A., Goldenberg, M.P., Hill, R., Jablonski, B.B.R., & Tropp, D. (2016) *The economics of local food systems: A toolkit to guide community discussions, assessments and choices.* Washington, DC, US Department of Agriculture Agricultural Marketing Service.

US Agricultural Marketing Service (AMS). US Department of Agriculture. Available at: https://www.ams.usda.gov/.

US Bureau of Economic Analysis (series). US Department of Commerce. Regional Economic Accounts. Available at: https://www.bea.gov/regional/index.htm.

US Bureau of Labor Statistics (series). US Department of Labor. Consumer Expenditure Survey. Available at: http://www.bls.gov/cex/home.htm.

US Center for Disease Control and Prevention (series). Behavioral Risk Factors Surveillance System (BRFSS). Available at: https://www.cdc.gov/brfss/index.html.

US Economic Research Service (ERS) (series). *Agricultural Productivity in the U.S.* US Department of Agriculture. Available at: https://www.ers.usda.gov/data-products/agricultural-productivity-in-the-us/.

US Economic Research Service (ERS) (series). *Agricultural Resource Management Survey (ARMS).* US Department of Agriculture. Available at: https://www.nass.usda.gov/Surveys/Guide_to_NASS_Surveys/Ag_Resource_Management/.

US Economic Research Service (ERS) (series) *Corn Production Costs and Returns per Planted Acre, excluding Government payments, 1996–2016. Data for the Heartland (central Midwestern states).* US Department of Agriculture. Available at: https://www.ers.usda.gov/data-products/commodity-costs-and-returns/.

US Economic Research Service (ERS) (series) *Farm Household Income and Characteristics.* US Department of Agriculture. Available at: https://www.ers.usda.gov/data-products/farm-household-income-and-characteristics/.

US Economic Research Service (ERS) (series) *Farm Income and Wealth Statistics.* US Department of Agriculture. Available at: http://ers.usda.gov/Data/FarmIncome/finfidmu.htm.

US Economic Research Service (ERS) (series). *Food Consumption Data.* US Department of Agriculture. Available at: http://www.ers.usda.gov/data/foodconsumption/.

US Federal Reserve Bank of Minneapolis (n.d.) *Consumer Price Index, 1913 – present.* Available at: https://www.minneapolisfed.org/community/financial-and-economic-education/cpi-calculator/consumer-price-index-and-inflation-rates-1913.

US National Agricultural Statistics Service (series). *Census of Agriculture.* US Department of Agriculture. Available at: http://www.nass.usda.gov/census/.

White, B. (2014) Comments to State of Indiana. Indiana Grown Commission. 17 September. Indianapolis, IN.

Williams, B. (2005) Soft systems methodology. Report for the Kellogg Foundation, Battle Creek, MI. Available at: www.bobwilliams.co.nz/ewExternalFiles/ssm.pdf.

Wiskerke, J.S.C. & Viljoen, A. (2012) Sustainable urban food provisioning: Challenges for scientists, policymakers, planners, and designers. In: Viljoen, A. & Wiskerke, J.S.C. (Eds.) *Sustainable food planning: Evolving theory and practice.* Wageningen, Netherlands, Wageningen Academic Publishers.

Part II

Operationalizing sustainable food system assessment

5 Data gaps and the politics of data

Generating appropriate data for food system assessment in Cape Town, South Africa

Jane Battersby

Introduction

There has been an upsurge in interest in governance of urban food systems[1] in the past decade. This has been typified by the rise of local food policy councils, as well as the emergence of national, regional, and global urban food governance networks. While governments and civil society have become increasingly aware of the need to proactively engage with urban food systems, it has become apparent that there are significant data gaps limiting good governance informed by historical framings of mandates of governments.

One of the themes emerging from trans-local governance networks has been the need for metrics and indicators to assess food systems. This need is particularly felt in the African context, where issues of data availability and reliability are a critical challenge for good governance generally (Borel-Saladin, 2017). These general data challenges are amplified in the case of data to inform food systems governance, as this has not been an area of focus of African governments or international agencies collecting and collating data (Haysom & Tawodzera, 2018).

At its heart, this chapter is a discussion on the relationship between the knowledge effects and governance effects of data and indicators (Prada Uribe, 2012). The ways in which the food security issue has been framed has shaped what data are gathered and how these are disaggregated and interpreted (knowledge effect). The data, in turn, reinforce the policy and programmatic focus of the state (governance effect). This chapter seeks to interrogate this mutually reinforcing relationship with regards to urban food governance and to propose new entry points that would enable the state to engage in new policy approaches.

To do so, this chapter reflects on efforts to generate a comprehensive assessment of the food system and the state of food security in Cape Town, South Africa, perhaps the most data resourced city on the African continent. The chapter highlights challenges experienced, suggests alternative approaches to data, and reflects on the potential impact of global data reporting processes, such as the UN Sustainable Development Goals (SDGs) and the 100 Resilient Cities programme on the ability of cities to effectively monitor and govern their food systems.

The City of Cape Town Food System and Food Security Study

In 2013 the City of Cape Town commissioned a Food System and Food Security study, co-funded by the City and the Provincial Department of Agriculture. The Terms of Reference articulated the need for the study thus,

> Food security or the lack thereof is the outcome of complex and multi-dimensional factors comprising a food system. Therefore, food insecurity is the result of failures or inefficiencies in one or more dimensions of the food system. This necessitates a holistic analysis of the food system that then can provide insights into the various components of the system, especially in our context as a developing world city. That analysis must also take note of the constitutional mandates of the tiers of government in South Africa, such as the legal mandate for food security that rests with the national government, in conjunction with various provincial departments. Local government, however, needs to understand food systems so as to make evidence-based planning and policy decisions that will have long-term impacts on their areas.
>
> (City of Cape Town, 2013, Tender number 414C/2012/13, in Battersby et al., 2014), p. 9)

This call therefore identified a need for the City to understand the nature of its food system as a means to address food insecurity. It further noted that the governance of food systems and food security is complex and requires cross-scale formal governance processes. This was a radical departure in the framing of food security by municipal governments in South Africa, and in Africa more broadly, and emerged out of the experiences of the City's Urban Agriculture Unit's efforts to implement its Urban Agriculture Policy (see also Chapter 4, this volume).

While the Terms of Reference articulated a wide-ranging interest in food systems, the subtext for the report was that the City of Cape Town was having a series of fraught political conversations within municipal government and with external stakeholders about the future of an area of agricultural land within Cape Town, the Philippi Horticultural Area (PHA). As the Mayor publicly articulated it, the report was commissioned to guide the City's decision-making (de Lille, 2013). According to a Council decision in 2012, no decisions were meant to be made about the future of the PHA until after the report had been completed (Davis, 2013). However, the City decided to override this moratorium on decisions before the report was completed, on the basis of the apparent urgent need for housing 'discovered' as a result of the release of 2011 census data (Lewis, 2013). What was already a highly politicized public debate became even more so, and questions of data and the role of data in governance came sharply into focus in the months that followed.

I was the leader of the team appointed by the City to conduct the study. The resulting 400 page report from the study identified roles of local government under existing constitutional mandates and programming foci, and provided an overview of the South African food system; Cape Town's food production of both commercial and small-scale urban agriculture; food flows into and within the city; retailing, processing, and sustainability; food price inflation; food insecurity; and lessons from elsewhere. The report concluded with mapping a potential governance approach with 31 recommendations all actionable within existing governance mandates of the City (Battersby et al., 2014).

The report was initially blocked by the City and then eventually released as a result of a Public Access to Information Act application by a food activist. When the City allowed the report to be released, it was on the condition that a Council Report on the study document be appended which stated,

> Whilst the study includes important contextual research, it has several limitation[s]. This is due, in part, to difficulties in accessing data, especially from private companies involved in food retail which make up the bulk of the food supply system. The lack of data meant that the study needed to rely on various assumptions in order to draw conclusions. As such, while it is informative background research, it will need to be considered with a number of other data sources, research and contextual factors not least of which are the forces of urbanization and the fact that food security, which has a bearing on food systems, is not a local government competency.
>
> (Battersby et al., 2014, no page)

This chapter uses the Cape Town process to reflect on questions of why data gaps exist in official data, what kinds of data are understood as useful to local government, and what possibilities exist for generating and consolidating data that would enable local government, and other tiers of government, to effectively engage in governance of the urban food system (see also Chapter 11, this volume).

Causes of the city-scale data gap and governance challenges

In their 2017 paper, Giordano et al. identify four clusters of challenges for embedding food systems governance in local governments in Africa. These are: lack of awareness, limited evidence at the local scale, incomplete decentralization processes, and financial challenges (Giordano et al., 2017, pp. 352–353). This chapter largely supports Giordano et al.'s framing of the governance challenge in African cities and seeks to illustrate how these factors manifest in the Cape Town context. However, this chapter also argues that their framing pays insufficient attention to the private sector's relative power in monitoring and shaping urban food systems in Africa (Battersby, 2017; Battersby & Muwowo, 2019).

This chapter argues that the challenges identified by Giordano et al. (2017) and the role of the private sector are all connected by a fundamental relationship between governance and data informed by deeply entrenched beliefs about the food system and food security. These beliefs are reinforced by data collection, aggregation and analysis decisions, which are in turn informed by the underlying beliefs. This sets up self-reinforcing feedback loops that entrench existing systems and make it exceptionally hard for new perspectives on food systems and food security to emerge and gain traction in policy.

Urban food systems governance challenges all have their origins in the historical and current framings of the food system and food security by national governments (see also Chapter 6, this volume). In the South African, and wider African context, food insecurity has been largely framed in policy as primarily rural and the responsibility of departments of agriculture (Crush & Riley, 2018). However, in their commissioning of the report, the City of Cape Town had come to understand that the problem of food insecurity is directly related to the food system, which has many urban components and determinants. The City was therefore interested in developing an understanding of what its mandates regarding food security were and, as a result, what role it could and should play in working towards food security. This required an analysis of mandates and the data used by the city to inform governance decisions.

Reinterpreting mandates

Within the 2014 report, we argued that City government has a wider mandate for food security than understood by government in general, particularly when acknowledging the food system as a key determinant of food security. Within the South African Constitution, the right to food is recognized (Section 27.1.b) and the state is obligated to ensure the progressive realization of this right (Republic of South Africa, 1996, p. 11). All spheres of government (national, provincial, and local) are bound by the Constitution and are required to respect, protect, and fulfil the rights guaranteed in the Constitution. Therefore, while the role of local government is not overtly articulated with respect to the right to food, it is still constitutionally bound to working towards the progressive realization of this right and to not undermine it.

The Food System and Food Security report argued that there are a number of powers and functions of local government outlined in the South African Constitution that directly or indirectly impact the food system and therefore food security. These include (as outlined in Schedules 4 and 5): licensing and control of undertakings that sell food to the public, local amenities, markets, municipal abattoirs, municipal parks and recreation, municipal planning, public places, refuse removal, and street trading. Furthermore Provincial and Local government have concurrent legislative competencies which impact food systems and food security, namely: agriculture, consumer protection, disaster management; education at all levels (excluding tertiary education), environment, health services, housing, industrial promotion, pollution control, population development, public transport, public works only in respect of the needs of provincial government departments in the discharge of their responsibilities to administer functions specifically assigned to them in terms of the Constitution or any other law, regional planning and development, soil conservation, trade, urban and rural development, and welfare services (Battersby et al., 2014, pp. 20–21; see also Chapter 11, this volume).

Within the report, we went on to examine the existing programmes and policies of each of the City of Cape Town's Directorates and their mandates in accordance with the national legislation guiding their actions. It found that all directorates already impacted food security – either positively or negatively. It further argued that according to existing mandates, local government does have a clear role to play in ensuring food security through urban food system planning and monitoring (see also Chapter 6, this volume).

However, in order for local government to play a role in food system and food security governance, it requires formal acknowledgement of this mandate – and therefore funding to follow through on obligations – and data to inform and support governance. This is a significant challenge, as the existing mandates as identified above do not explicitly mention food or food security, and the national policy framework's framing of food security omit explicit identification of local governments as agents in food security. Most importantly, the national government's 2002 Integrated Food Security Strategy set out an argument that food security was built on a robust food system, and to achieve food security it was essential to have a multi-departmental approach. However, in reality, implementation of the strategy fell to the Department of Agriculture (now the Department of Agriculture, Forestry and Fisheries). This had two deleterious impacts for urban food system and food security governance.

First, it reinforced an historical framing of food security as a problem to be addressed primarily through production-oriented policies and programmes (Drimie & Ruysenaar, 2010). Second, through placing the strategy under Agriculture, it effectively made food security a provincial government concern, as within South African government structures Agriculture is a provincial rather than local government department. This institutional location of the food security strategy signalled to local government that it did not have a clear role in food security. The impact of this framing was that local government food system- and food security-related mandates established within the constitution are poorly acknowledged by local, provincial, and national government. It also means that the kinds of data collected to measure and monitor food security are poorly designed and poorly disaggregated, to aid local government in making food system policy and programme decisions.

In 2014, the government released its National Policy on Food and Nutrition Security which was meant to provide a broad framework for the fulfilment of the Constitutional mandates for food security and to 'serve as a guide to national, provincial and local government in working towards food and nutrition security at every level' (DAFF, 2014, p. 29). However, as with the Integrated Food Security Strategy, the implementing departments operate at national and provincial scales, and have no local government representation. The challenges of framing and data therefore persist.

The consequences of the national government framing of food security has been that although key areas of the food system that shape urban food security are under local government mandates (most directly markets, street trading, land-use planning, and municipal abattoirs), local governments have not considered themselves to have food security or food system roles. Further, the framing of food insecurity as a primarily rural problem to be solved by production-based solutions has meant that there are significant gaps in official statistics that could guide local government decisions on food systems and food security.

The following sections highlight the kinds of data challenges impeding local government from developing informed food system and food security policies and programmes. There are four major categories of data challenges: poor disaggregation; dependence on weak proxies; a lack of relevant indicators for local government data sets; and private sector control of data. The lack of awareness and incomplete decentralization identified by Giordano et al. (2017) have fed into these challenges, which lead to limited evidence at the local scale, which in turn reinforces the lack of awareness and reduces the apparent urgency of decentralization.

Data challenge 1: poor disaggregation

Data are never neutral. They are collected to answer particular questions deemed important to the agency collecting them. In the case of official statistics, they are collected and collated to help governments understand and respond to policy issues. In the case of both food security and food systems data, the way in which they have been collected and collated reflects and reinforces a particular understanding of food security. This section of the chapter pays attention to issues of disaggregation and interpretation of disaggregated data.

Food security data

As noted above, food insecurity has been framed in policy documents as a primarily rural challenge in the South African context. This links to a notion that poverty is primarily rural, and that food insecurity and income poverty are directly correlated. However, it has been argued that the perception of poverty as rural is the outcome, in part, of data disaggregation. In 2003, StatsSA, South Africa's national Statistical Service, produced a 187-page discussion document addressing the challenge of finding an appropriate definition for 'urban' and 'rural' (StatsSA, 2003). Although this may seem excessively long, it is important because the continued use of apartheid era definitions of urban has inflated the country's rural population figure. The use of these apartheid era definitions has meant that many areas that would be considered to be urban using any standard definition, are considered rural. Many of these areas are poor. By classing these areas as rural, the myth of poverty being a rural problem is perpetuated, as is the interpretation that black African people in South Africa are predominantly rural (and live rural lives) (Parnell, 2005).

Furthermore, the state often falls into using percentages of populations instead of absolute numbers to make policy arguments. For example, in their 2015–2020 Strategic Plan, the Western Cape Provincial Department of Social Development argued that provincial focus on food security should centre on rural areas because

> According to the 2013 GHS [General Household Survey], 16.1% of households in the Western Cape have inadequate access to food, while 6.6% have severe inadequate access to food. In total, 22.7% of households are food insecure. Food insecurity is more prevalent in rural areas, where 27% of the population have inadequate access to food. The corresponding figure for urban areas is 20%.
>
> (Western Cape Government Social Development, 2015, p. 8)

However, approximately 90 per cent of the province is urban (with approximately two-thirds of the population residing in Cape Town). When these proportions are translated in numbers of households, there are 44,118 food insecure rural households, on a total of 294,120 food insecure households. Similar misuse of proportions over absolute numbers is found in the 2002 Integrated Food Security Strategy, which framed national food security priority areas (Battersby et al., 2014, p. 28). Poorly disaggregated data compounded with poor use of statistical data have reinforced the perception that food insecurity is not a significant urban challenge and therefore does not require specifically urban funding or programming.

Food system data

The challenge of disaggregation is further evident in food system data. One of the primary reasons for the City's commissioning of the Food System and Food Security Study was to develop an understanding of how much food was being produced in and around the city, and the importance of local production to local consumption (see also Chapter 10).

However, as with the food security data, the official statistics reflect historical assumptions about food security and its connection to the food system. Historically, interest in the food system in South Africa has focussed on issues of aggregate production. This was a particularly important question during the apartheid era, when the nation needed to have high self-sufficiency as a result of political and economic isolation. At that time, there was considerable support for commercial agriculture, and a sophisticated system of marketing boards and subsidies to maintain agricultural production. These boards controlled 70–80 per cent of the marketing of agricultural products. Within this context, the urban food system was subject to very little formal governance, outside of local government's management of fresh produce markets and abattoirs and their regulation of informal trade.

Stats-SA produces an Agricultural Census of Commercial Agriculture every ten years, the most recent of which was released in 2011 using data from 2006 and 2007. This report provides data on the 'production performance of all farmers who responded to the census' (StatsSA, n.d.).

The Agricultural Census had two critical flaws that made it hard to gauge how much food was being produced in and around Cape Town. The first is the partial nature of the data set. The Census collects data only from commercial agriculture, thereby discounting smallholder and subsistence agriculture. It also only collects data on a limited number of key crops: wheat, potatoes, onions, carrots, cabbages, apples and pears, and wine grapes; and livestock: cattle, sheep, pigs, ostriches, and chickens. There are significantly more crops being grown in Cape Town than these. Within our 2012 report on the PHA alone, we found that over 50 different crops were being produced (Battersby-Lennard & Haysom, 2012, p. 8). The main crops by volume were: cabbage, lettuce, cauliflower, broccoli, spinach, carrots, potatoes, and onions (Battersby-Lennard & Haysom, 2012, p. 41). Only four of these are reflected in the Agricultural Census. Additionally, the area is an important source of high value, highly perishable crops, such as herbs, for Cape Town.

The official production data on agricultural produce is framed on a particular vision of the agricultural sector and its value in the South African economy (similar challenges for indigenous communities are presented in Chapter 3, this volume). The limited range of crops monitored therefore obscure the real productive capacity of the land and therefore the land's value as an agricultural resource.

The second challenge was that the data were inappropriately disaggregated for municipal government use. The Agricultural Census disaggregates production data at the Magisterial District level (for reasons that are not entirely clear, since magisterial districts were designed to align judicial service boundaries). Nine different magisterial areas are found within the City of Cape Town's borders, but some of these extend beyond municipal boundaries. They do not align to other demarcations used by the municipality to govern or collate data, such as Ward boundaries, sub-district boundaries or health districts. In 2008, the City of Cape Town conducted an Agricultural Land Review assessing the value of the agricultural lands in and around the city. This report identified 13 different productive areas. The boundaries of magisterial districts and of these productive areas are unrelated. It was therefore impossible to use official statistics generated by national government to report on how much food was being produced in Cape Town and which of the agricultural areas identified by the City were the most productive.

Data challenge 2: weak proxies

Historically, urban governments have not been mandated to collect data on their food systems, as neither food systems nor food security governance has been recognized as part of their competencies. This leaves them with significant data gaps when they attempt to engage with food systems. They therefore depend on proxy data to build a narrative from contingent data that were collected for other reporting purposes.

Building on the broader question of how much food was being produced in Cape Town, the City had a specific interest in the role of the PHA as part of Cape Town's food system. As noted above, the future of the PHA has been hotly contested over the last ten years and has been a topic of considerable public debate.

The City had come to believe that the PHA's productivity levels were declining on the basis of two proxy indicators: volumes of produce entering the Cape Town Fresh Produce Market from the PHA, and the number of farmers operating in the PHA. Both proxy indicators provide inaccurate assessments of the PHA's productivity, and their use was informed by poor understanding of the South African food system.

The Cape Town Fresh Produce Market is the city's primary fresh produce market and was a municipal facility until it was privatized in 2004. Local governments have the constitutional mandate to manage municipal markets, and so historically have held data on market throughput and pricing. The City government has used the market's figures, which indicated reduced flow of produce from the PHA to the market, to infer declining importance of the PHA as a source of food for Cape Town (Battersby et al., 2014, p. 150). However, the City failed to appreciate the rapid changes in the South Africa food system since the end of apartheid. As a result of market deregulation, supermarkets play an increasingly important role in the food system. Farmers in the PHA estimate that they now sell 80 per cent of their produce directly to retailers and just 12 per cent now goes to the fresh produce market (Battersby et al., 2014, p. 98). Using fresh produce markets to infer production rates is fundamentally flawed.

The second indicator used is the number of farmers active in the PHA. The City inferred that decline in the number of farmers active in the PHA, by about half from 1994 to present, indicated declining production. However, while the number of farmers has declined, the land under production has actually increased as farms have been consolidated in the past 25 years (Battersby-Lennard & Haysom, 2012, p. 38). In addition, the rate of decline in farmer numbers in the PHA is actually well below the national average. Between 1990 and 2008, the number of farmers in South Africa declined by a staggering 76 per cent (Vink & Van Rooyen, 2009). As in the PHA, while the number of farming units has declined, national productivity has not. Farm consolidation is a national phenomenon and not an indicator of declining agricultural production.

Due to local government's historic and current lack of engagement with national issues of food and farming, the City has made inferences about the productive value of the PHA and its contribution to Cape Town's food system, informed by local proxy data used without an appreciation of the national context of these data.

Data challenge 3: local government data sets lack relevant indicators

Given the historical lack of interest in food system issues, the Food System and Food Security Study found very little data generated by the City that could assist them in understanding the nature of the food system and make informed policy decisions.

In one case, it appeared that City officials failed to understand the limitations of the data they were drawing on to support decisions. In other cases, it was clear that the data gaps were understood, but under existing conditions of perceived mandates and associated funding it was not possible to extend the data collected on the various components of the food system.

Failure to appreciate limitations of data collation

There is significant pressure from developers to convert agriculturally zoned land to an urban zoning to enable development, despite the City's own report in 2010 indicated that there is enough land for all required urban development within the demarcated Urban Edge until 2021 (City of Cape Town, 2010, p. 9). The City therefore established the following principle to protect agricultural land for food security purposes in its 2012 Spatial Development Framework.

> To promote food security and mitigate food price increases, the City should therefore consider having 'high-potential and unique agricultural areas'... declared as agricultural/cultural landscapes by the highest appropriate level of authority; investigate ways in which all agricultural areas of significant value ... could receive local protection (over and above the urban edge). Options include environmental or heritage overlay zones applied through the relevant zoning regulations; ... inside and outside the urban edge proactively prepare and implement action/

> management plans that prevent encroachment and unlawful land use in agricultural areas, minimize negative impacts of urban development on farmed land and manage the use of water and other natural resources.
>
> (City of Cape Town, 2012, p. 65,

The Spatial Development Framework is a long-term (20+ year) city-scale plan to guide new growth and change in the City, and is meant to guide all area-based planning.

However, the Food System and Food Security study found that it was not possible to apply this principle in practice because of how agricultural land valuations had been conducted. In 2008, the City commissioned an Agricultural Land Review, which assessed the relative value of 13 agricultural areas in Cape Town. This report was conducted by an agricultural economist and was informed by a

> socio-economic empowerment role in terms of food production, food security and contribution to LED [local economic development]; its economic role in food production and other commodities (e.g. Wine), especially as input to the secondary and tertiary industry; and its relationship to the City's green structure and biodiversity corridors.
>
> (City of Cape Town, 2008, p. 1)

The assessment was made on the basis of five main criteria: agricultural potential, economic significance, land-use significance, landscape significance, and environmental significance. Each criteria had a number of subcategories against which their value was assessed and ascribed a value of 'low', 'medium' or 'high' by a team of agricultural specialists. This subjective approach, informed by a very particular set of ways of valuing agricultural land, identified three agricultural areas classified as 'high potential and unique', namely, Constantia Hills, Helderberg/Erinvale, and Philadelphia. The PHA was noted as the fourth most valuable, but importantly did not fall into the 'high potential and unique' category. It is notable that the top two areas are primarily viticulture areas and so cannot be considered as areas that promote food security and mitigate food prices, as articulated in the Spatial Development Framework (SDF). The cumulative impact of the Agricultural Land Review approach is that certain agricultural areas are argued to be of global importance and thus irreplaceable (viticulture), and that areas producing vegetables and grains, which contribute more directly to food security, are not.

The officials who had drafted the SDF had not appreciated the ways in which the food system and its value had been framed by the earlier assessment and therefore were attempting to use it for purposes beyond what it was intended for and then ultimately undermined their objective. We argued with the Food System and Food Security Study that it would be of value for the City to commission a new Agricultural Land Review informed by the food security contribution of productive lands. It is also essential that municipal officials contributing to strategic planning documents are better informed about what data have been used to inform reports subsequently drawn on to make planning decisions.

Absent data

In addition to the questions of the significance and value of productive land, the Food System and Food Security Study also sought to describe the food system as a whole and identify its points of resilience and vulnerability. This was challenging for a number of data reasons.

As noted earlier, there had previously been no consideration of the nature and dynamics of the urban food system in South African governance processes. As well, food systems governance under apartheid was characterized by support for large-scale commercial agriculture, the dominance of marketing boards, and cooperatives with state-appointed boards. The urban food system was subject to very little formal governance, outside of local government's management of fresh produce markets and abattoirs and their regulation of informal trade. As a result of this absence of formal governance, little data was gathered on components of the urban food system. Following the end of apartheid, the marketing boards were dissolved, and the state stopped monitoring and regulating food flows.

Local and provincial governments only collect data expressly related to their abilities to monitor and report on their core mandates. This is entirely appropriate and a good use of limited municipal finances. However, it does mean that official data are not held on important aspects of the food systems operating in cities.

Both city and provincial governments collect data on urban agriculture projects supported by government. However, the data held by the provincial government were limited to data on physical location and number of beneficiaries. Production data did not extend beyond broad categories of 'vegetables' or 'chicken'. The data supplied by the City to our team as the available data on supported projects were handwritten delivery instructions for compost and manure. There appeared to be no consolidated database even of addresses of supported gardens. Due to the absence of a food security mandate, urban agriculture was housed within the Economic Development Department and therefore viewed as a development initiative and a means of livelihood support. As such, the focus of monitoring and evaluation is strongly input based. This input-based evaluation approach is further necessitated by extremely limited personnel and the financial capacity of the state to engage in output-based evaluation. In conversation with a representative from the Provincial Farmer Support and Development programme, I asked why data were not collected on farm outputs. He explained it was too difficult to get farmers to report on production. I then asked how they knew if a project was successful and sustainable. He answered that if the beneficiaries asked for resources again the next year, the project was sustainable. In other words, a project was sustainable if it continued to rely on government support.

In terms of food processing, the City of Cape Town holds a database of licensed food processors within the city. However, this database's purpose is simply to guide environmental health inspections. It therefore identifies the physical location of the business and the broad category of food stuffs being processed. It provides no information on the scale of the business, the source of the raw materials used, or the destination of the product.

Within Cape Town, informal vendors trading in food need to apply for a 'hawking in meals' licence and obtain a 'certificate of acceptability'. However, despite these regulations, the City does not hold any data on informal food retail (what is being sold). These data are simply held to allow the City to conduct period health and safety checks according to its environmental health mandate. Additionally, although understood to be a major source of livelihoods in the city, representatives from City government on the steering committee of the Food System and Food Security Study viewed the informal sector to be quite marginal to the food system, despite the fact that the informal sector accounts for at least 25 per cent of food retail in South Africa (Agyenim-Boateng et al., 2015), and is particularly important for low-income consumers (Cooke, 2012).

Finally, there has been a rapid expansion of supermarkets and associated fast food outlets in Cape Town, which have rapidly transformed the food system (Battersby, 2017). The Planning and Building Development Management department are responsible for allocating development rights for new retail developments in the city, but their decision-making process does not consider the food system impact of planning decisions and therefore do not monitor the number of formal food retailers.

The outcome of all of this is that although many components of the food system fall under the governance of local government, and local government collect data on these components, the data are limited to addressing issues that are understood to be local government's core mandate. It therefore leaves the State unable to effectively monitor and evaluate its food system and its relationship to food security, or to justify the development of programmes or policies for food system governance.

Data challenge 4: private sector control of data

The final data challenge is that data pertaining to the food system are increasingly controlled by large-scale private sector actors through the vertical coordination of the supply chain and the increasing dominance of large supermarkets within the formal market. Post-apartheid deregulation of the food system has made it increasingly opaque. The consolidation and concentration in the food sector have made the food system less traceable, as large companies are unwilling to share their data. As Isabel Schmidt from StatsSA observed:

> Statistics South Africa has noted that the abolition of marketing and control boards in the food sector has resulted in significant information gaps in relation to food systems in South Africa. Furthermore, the activities of the Competition Commission [are] increasingly making non-regulated associations and large scale conglomerates unwilling to voluntarily provide information about their activities to the Agricultural marketing board and other state entities.
>
> (Schmidt, StatsSA & Pers. Comm. in Battersby et al., 2014, p. 67)

The private sector control of data and lack of accountability has been acknowledged to have caused significant problems for food security and food systems governance. The most telling of these has been the listeria outbreak, which resulted in over 200 deaths in South Africa in 2017 and 2018 (Marler, 2018). Two factories producing polony (a ready-to-eat meat product) for Tiger Brands were eventually identified as the source of the listeria outbreak. In South Africa, the ten largest packaged food companies account for around 52 per cent of sale of packaged foods, with Tiger Brands being by far the largest company, having over 17 per cent of the market share of all packaged food sales in the country (Igumbor et al., 2012, pp. 2–3). The combination of this market concentration with the rolling back of state control of the food system has reduced traceability and regulatory power. In the wake of the listeria outbreak, it emerged that the processed meat industry had blocked the implementation of regulatory standards developed in 2014 (Ensor, 2018). In addition to the power ceded to large private-sector actors, there has been an erosion in the capacity of the state to monitor the food system within its existing mandates, given funding shortfalls. At the height of the listeria outbreak, the Minister of Health said, 'We have a serious shortage of environmental health inspectors because that function was given to local government in 1996 … It was a mistake for the constitution to give that job to municipalities because they cannot do it.' He further noted that there was a shortage of 3300 municipal environmental health officers nationally (Mkhwanazi & African News Agency, 2018).

This has therefore led to a position in which industries are effectively left to self-monitor, as the limited state capacity is generally focussed on regulating the informal sector. The long-term trend has therefore been towards greater corporate power and diminished state capacity to regulate the food sector.

Moving forward

When the City commissioned the Food System and Food Security Study, it took a bold move, being the first metropolitan area to seek to engage in the food system in a holistic manner and attempt to understand what role the city needs to play in the food system. The report argued that local government could be understood to have a significant mandate to govern the food system and food security. However, moves towards acting upon this revealed that the mandate was hampered by the web of knowledge and governance effects constructed by historical framings of food security by national government. This inertia was amplified by a lack of political will and vested interests. Echoing Giordano et al.'s (2017) framing, the lack of complete decentralization has impacted upon funding and capacity at the local scale, which further amplifies the inertia.

The increased interest in governance of urban food systems around the world has highlighted the importance of new forms of governance informed by new forms of data. In the absence of official statistics suitably conceived and disaggregated to address urban food governance questions, there is a necessity for local government to locate or generate appropriate data. Given the connections between food security, food systems, and urban systems, there is a need for data that examine this nexus. The International Institute for Environment and Development has done innovative work in Nairobi, Kenya, examining urban food security's connection to the local urban food system and urban form (Ahmed et al., 2015; see also Chapter 8, this volume). In order to understand the interactions of food systems, urban systems, and food security, it will be essential to start from the most vulnerable to food insecurity and work upwards and outwards towards a systemic understanding. This therefore requires a move from large-scale official data sets towards micro-scale, collaborative, community-generated data and assessment frameworks.

However, 'doing data' in different ways has economic costs and is potentially politically risky as local governments are concerned about being seen to act beyond their mandates. Furthermore, increasingly local governments in Africa are being led by externally generated data imperatives. Governments are currently particularly focussed on reporting against UN Sustainable Development Goals indicators. In the context of urban food systems governance, this is very concerning. The absence of the urban in the hunger/food goal (SDG 2) and the corresponding absence of food in the urban goal (SDG 11) limits any proactive urban food security or food system data collection at the urban scale (Haysom & Tawodzera, 2018, p. 23). Cape Town is a member of the 100 Resilient Cities network and therefore is reporting against their indicators. There is concern that these indicators are poorly aligned to African urban realities and may lead the City along development pathways that undermine food system resilience rather than enhance it. There has been some push back to indicator sets that are poorly aligned to realities of the Global South from some members of the 100 Resilient Cities network, and a number of studies have been conducted examining the viability and utility of the indicators within the various Sustainable Development Goals. It is essential that local and national governments engage critically with externally produced indicators and provincialize them to meet existing local concerns. This, however, depends on significant local capacity. In the absence of such capacity, this chapter, along with the Food System and Food Security Study argue that the state should draw on the capacity of academic institutions to address these challenges.

If Cape Town and other cities are to work towards proactive urban food systems governance, it is going to be essential for local government to understand the inherent politics of measurement. This will require three steps going forward: (1) understanding why the data have gaps; (2) understanding both what the existing data reveal and what they obscure and why; and (3) identifying what data are required to help local governments address their full mandates regarding food. Only through these steps will it be possible for local government to unpick the Gordian Knot of governance and knowledge effects of data and indicators.

Acknowledgements

I would like to acknowledge the conversations with Gareth Haysom of the *African Centre for Cities* in honing a number of the arguments made in this chapter.

Note

1 Although there has been an upsurge in interest in food systems it is important to note that the precise meaning of the term 'food systems' varies considerably and the term has been mobilized by researchers and policymakers for different purposes. For the purpose of this chapter, I use the following definition,

> A food system gathers all the elements (environment, people, inputs, processes, infrastructures, institutions, etc.) and activities that relate to the production, processing, distribution, preparation and consumption of food, and the outputs of these activities, including socioeconomic and environmental outcomes.
>
> (HLPE, 2014, p. 12)

References

Agyenim-Boateng, Y., Benson-Armer, R., & Russo, B. (2015) *Winning Africa's Consumer Market*. New York, McKinsey & Company. Available at: http:// www.mckinsey.com/industries/consumer-packaged-goods/our-insights/winning-in-africas-consumer-market.

Ahmed, S., Simiyu, E., Githiri, G., Sverdlik, A., & Mbaka, S. (2015) *Cooking up a storm: Community-led mapping and advocacy with food vendors in Nairobi's informal settlements*. International Institute for Environment & Development (IIED) Working Paper, London, IIED.

Battersby, J. (2017) Food systems transformation in the absence of food system planning: The case of supermarket and shopping mall expansion in Cape Town, South Africa. *Built Environment*. 43(3), pp. 417–430.

Battersby, J. & Muwowo, F. (2019) Unacknowledged food systems governance in Kitwe, Zambia. In: Battersby, J. & Watson, V. (Eds.) *Urban food systems governance and poverty in Africa*. London, Routledge, pp. 128–140.

Battersby, J., Haysom, G., Tawodzera, G., McLachlan, M., & Crush, J. (2014) *Food system and food security study for the City of Cape Town*. Available at: http://www.afsun.org/wp-content/uploads/2016/08/Final-Food-System-Study-Report_Corrected-_WITH-COUNCIL-REPORT.pdf.

Battersby-Lennard, J. & Haysom, G. (2012) *Philippi horticultural area: A city asset or potential development node?* Report by the African Food Security Urban Network, African Centre for Cities, University of Cape Town.

Borel-Saladin, J. (2017) Where to draw the line: Data problems and other difficulties estimating urbanisation in Africa. *Development Southern Africa*. 34(2), pp. 137–150.

City of Cape Town (2008) *Agricultural land review*. Report 1498/2, compiled by iKapa Enviroplan. City of Cape Town, October.

City of Cape Town (2010) *Evaluation of developable land within urban edge*. City of Cape Town, Cape Town.

City of Cape Town (2012) *Cape Town spatial development framework: Statutory report*. City of Cape Town, Cape Town.

Cooke, K. (2012) *Urban food access: A study of the lived experience of food access within a low income community in Cape Town*. Unpublished MA Thesis. University of Cape Town, Cape Town.

Crush, J. & Riley, L. (2018) Rural bias and urban food security. In: Battersby, J. & Watson, V. (Eds.) *Urban food systems governance and poverty in African cities*. London, Routledge, pp. 42–55.

DAFF (Department of Agriculture Forestry and Fisheries) (2014) *The National Policy on Food and Nutrition Security 2014*. Pretoria, Government of South Africa.

Davis, R. (2013) The battle for Cape Town's farmlands. *Daily Maverick*, 1 August. Available at: https://www.dailymaverick.co.za/article/2013-08-01-the-battle-for-cape-towns-farmland/.

De Lille, P. (2013) *The proposal to move the urban edge does not compromise productive capacity of the PHA*. Somerset West Ward 84, Cape Town, South Africa, 29 July. Available at: https://www.facebook.com/notes/patricia-de-lille/the-proposal-to-move-the-urban-edge-does-not-compromise-productive-capacity-of-t/581552041883862/.

Drimie, S. & Ruysenaar, S. (2010) The Integrated Food Security Strategy of South Africa: An institutional analysis. *Agrekon*. 49(3), pp. 316–337.

Ensor, L. (2018) Before listeria: How the processed-meat industry 'blocked' regulation. *Sunday Times*, 9 March. Available at: https://www.timeslive.co.za/sunday-times/business/2018-03-09-before-listeria-how-the-processed-meat-industry-blocked-regulation/.

Giordano, T., Taylor, K., & Touadi, J.L. (2017) Urban food policies – sharing food system innovations through decentralized cooperation in Africa. *JUNCO- Journal of Universities and international development Cooperation*. 1, pp. 348–356.

Haysom, G. & Tawodzera, G. (2018) 'Measurement drives diagnosis and response': Gaps in transferring food security assessment to the urban scale. *Food Policy*. 74, pp. 117–125.

HLPE [High Level Panel of Experts] (2014). *Food losses and waste in the context of sustainable food systems*. A report by the High-Level Panel of Experts on Food Security and Nutrition of the Committee on World Food Security. Rome, HLPE.

Igumbor, E., Sanders, D., Puoane, T., Tsolekile, L., Schwarz, C., Purdy, C., Swart, R., Durão S., & Hawkes, C. (2012) 'Big Food,' the consumer food environment, health, and the policy response in South Africa. *PLoS Medicine*. 9(7).

Lewis, A. (2013) Urban edge: City wants to develop crucial Philippi Farmlands. *Cape Argus*, 17 July. Available at: https://www.pressreader.com/south-africa/cape-argus/20130717/282548720875133.

Marler, B. (2018) 216 Dead with 1,060 with Listeria in South Africa – People still getting sick from polony. *Food Poison Journal*, 2 August. Online. Available at: https://www.foodpoisonjournal.com/food-poisoning-information/216-dead-with-1060-with-listeria-in-south-africa-people-still-getting-sick-from-polony/.

Mkhwanazi, S. & African News Agency (2018) #Listeriosis: Minister pledges support for families of dead who take on companies. *Cape Times*, 9 March. Available at: https://www.iol.co.za/capetimes/news/listeriosis-minister-pledges-support-for-families-of-dead-who-take-on-companies-13669378.

Parnell, S. (2005) Constructing a developmental nation: The challenge of including the poor in the post-apartheid city. *Transformation: Critical Perspectives on Southern Africa*. 58, pp. 20–44.

Prada-Uribe, M.A. (2012) *Development through data? A case study on the World Bank's performance indicators and their impact on development in the Global South*. IRPA Research Paper No. 5. Available at: http://dx.doi.org/10.2139/ssrn.2167366.

Republic of South Africa (1996) *Constitution of the Republic of South Africa*. Act 108 of 1996. Available at: www.info.gov.za/documents/constitution/1996/a108-96.pdf.

StatsSA (2003) *Census 2001: Concepts and Definitions*. Pretoria, Statistics South Africa.

StatsSA (n.d.) About: Report-11-02-02 – Census of Agriculture Provincial Statistics. Financial and production statistics, Western Cape. Available at: http://www.statssa.gov.za/?page_id=1861&PPN=Report-11-02-02&SCH=5122.

Vink, N. & Van Rooyen, J. (2009) *The economic performance of agriculture in South Africa since 1994: Implications for food security*. Development Planning Division, Development Bank of Southern Africa, Midrand, South Africa.

Western Cape Government: Social Development (2015) Strategic Plan 2015–2020. Available at: https://www.westerncape.gov.za/assets/departments/social-development/strategic_plan_2015-20.pdf.

6 Action research as a tool to measure progress in sustainable food cities

Enacting reflexive governance principles to develop indicators

Ana Moragues-Faus

Introduction

In the context of rapid urbanization, increasing socio-economic inequality and accelerating ecological degradation (Godfray et al., 2010; OECD, 2015), cities are recognizing the urgent need to transform food systems to deliver health and well-being, prosperity and equity, and conserve and enhance natural resources. Indeed, an increasing number of cities around the world are now adopting collaborative and cross-sector approaches that use food as a primary vehicle for delivering positive social, economic, and environmental outcomes (Moragues-Faus, 2017d; see also Chapters 5, 8, 9, 10, & 11, this volume).

In the last few years, there has been an increasing interest by practitioners and academics alike to measure the impact of urban food policies and associated initiatives. Recent exercises include a mixture of local, national, and international processes that aim to measure progress in individual cities around the delivery of sustainability and food security outcomes (Tanguay et al., 2010; Sustainable Development Solutions Network, 2015; RUAF, 2016; Landert et al., 2017). However, food system assessments face a number of challenges to become effective tools for food system transformation, from the need to critically examine how knowledge is produced and the conflicting interests between stakeholders, to the lack of data or the multi-scalar usefulness of specific indicators (Alrøe et al., 2016; Carlsson et al., 2017; also see Chapter 6, this volume). To address some of these challenges, this chapter explores the potential role of action research (AR) – and particularly co-productive and reflective practices – in developing these measurements (see also Chapters 2, 3, 4, & 10, this volume). For that purpose, I rely on an action research process conducted by Cardiff University and the Sustainable Food Cities network – an initiative made up of 55 urban food partnerships across the UK. The research process involved more than 100 sector experts and practitioners to develop a place-based indicators toolbox to measure progress and inspire action.

The remainder of the chapter is structured as follows. The second section presents a literature review on food system assessments in order to identify key challenges in the field and discuss how action research processes can contribute to address key concerns. The third section discusses the action research process and its main outputs. The fourth section analyses how co-productive and reflective practices were implemented throughout the research processes and resulted in a flexible, participative, horizontal, and place-based approach to develop food system assessments. The analysis leads to an examination of the benefits and limitations of this particular research process. Finally, the chapter concludes by highlighting how critical and reflexive governance principles can be applied to the development of food assessments as well as other research processes to progress conceptual debates and the material construction of more sustainable food systems (see also Chapters 5 & 11, this volume).

Literature review on food system assessments: towards participatory processes

There is an ever-increasing number of exercises evaluating the performance of food systems. A recent review of sustainability assessments in the food sector revealed the diversity of purposes that these tools serve, including: conducting research and collecting data; providing advice to different actors such as policymakers, consumers, or farmers; monitoring specific activities such as farmers' certification, self-assessment, or landscape planning (Schader et al., 2014). In many instances, these exercises focus on one dimension of sustainability – such as studies on planetary boundaries (Conijn et al., 2018), or one aspect of the food chain – for example, organic agriculture (Underwood et al., 2011). However, in this research process it was particularly important to take a holistic approach to food system reform, which embraces the horizontal and vertical dimensions of the food system (Moragues-Faus et al., 2013). Horizontally, a holistic food system includes different policy domains such as health and well-being, environment, economy and community development, social and cultural aspects, and education. The vertical dimension refers to all stages of the food system from food production, processing and storage to transport, retail, consumption, and waste.

While sustainability looms large as the main focus of food assessments (see FAO, 2014; Dubbeling et al., 2017), other concepts are also being mobilized that similarly adopt a holistic food system perspective, such as resilience of the global food system (Seekell et al., 2017), sustainable nutrition security (Gustafson et al., 2016), agroecology (Hatt et al., 2016), sustainable intensification (Mahon et al., 2017), food security (Ingram, 2011) or sustainable diets (Johnston et al., 2014). Many of these exercises highlight key areas of concern or dimensions that food assessments need to incorporate in their analysis. Nevertheless, few of these more holistic accounts of food systems actually provide specific indicators to guide data collection (see exceptions such as FAO, 2014; Gustafson et al., 2016).

Holistic evaluations of urban food system assessments are still scarce (Prosperi, Moragues-Faus et al., 2015). The separation of the dimensions of sustainable urban development hinders the ability of most systems to accurately understand broad movements towards sustainability (Lynch et al., 2011). To date, an important part of the urban food literature has concentrated on evaluating specific activities, such as the contribution of urban agriculture to sustainability (Hanson & Schrader, 2014). However, recent works are taking a more holistic approach to evaluating progress in building more sustainable urban food systems (see, for example, Ilieva, 2017; Landert et al., 2017). For instance, efforts have been directed towards assessing and planning city region food systems (Dubbeling et al., 2017). As part of this process, the International Resource Centre on Food Security and Agriculture (RUAF), the Food and Agriculture Organisation (FAO), and Wilfrid Laurier University have developed an indicator framework to assess and monitor different dimensions and goals of a sustainable city region food system.[1] As this exercise and the case study presented here show, practitioners and international institutions are driving the need for these exercises. For example, the Milan Food Policy Pact Secretariat – a protocol signed by more than 160 city Mayors across the globe to build sustainable urban food systems – is also developing a framework to monitor progress in signature cities.[2]

There has been a clear evolution not only in the content of food system assessments towards more holistic understandings of sustainability dynamics, but also in reformulating the purpose and processes inherent to evaluating food systems. However, more critical reflection is needed on how knowledge is created, codified, and used in sustainability assessment exercises. The literature reveals a set of challenges in conducting food system assessments (see (Hiremath et al., 2013; Alrøe et al., 2016, 2017; Carlsson et al., 2017), mainly:

1. *How to balance different types of knowledge.* This is particularly relevant first, to avoid the most widely used or easiest ways to measure dimensions of sustainability getting the most weight; and second, to include different knowledges and experiences when representing contested sustainability views (Moragues-Faus & Marsden, 2017).
2. *How to expose the values mobilized through assessment exercises and related choices.* In order to actually address sustainability criteria, exercises should reflect on how they relate to the ethical principles of sustainable food systems as well as how they deal with the interests of different stakeholders (Moragues-Faus et al., 2017). Working through sustainable food challenges and trade-offs necessarily requires addressing not only the power imbalances configuring the contemporary food system, but also those at play in developing assessment frameworks. Examples range from who is invited, who has the time and resources to participate, or what type of knowledges/experiences are

regarded as legitimate (i.e. those based on 'scientific' evidence). Taking a participatory perspective, from problem definition to the identification of goals and the development of solutions, can be instrumental in exposing these variegated interests and capacities at play (Travers, 1997; lrøe et al., 2017), as well as providing a platform to voice the needs and lived experiences of those relegated to the margins of the food system.

3. *How to develop, implement, and communicate food system assessments in a way that effectively contributes to the development of more sustainable food systems.* Among others, food system assessment exercises should be relevant for practitioners and local communities and allow for capturing specific place-based relations between different social, economic, and environmental sustainability dimensions. Nevertheless, connecting these local specificities to regional and global processes constitutes a key step towards representing the multilevel dynamics resulting in different sustainability and food security outcomes across the globe.

Action research and participatory processes can be instrumental in overcoming some of these challenges. Action research is broadly defined as

> a participatory, democratic process concerned with developing practical knowing in the pursuit of worthwhile human purposes ... It seeks to bring together action and reflection, theory and practice, in participation with others, in the pursuit of practical solutions to issues of pressing concern to people, and more generally the flourishing of individual persons and their communities.
>
> (Reason & Bradbury, 2008, p. 2)

In this manuscript, we focus on two key concepts inherent to action research processes: co-production and reflective practices.

First, the notion of *co-production* has become a key word in critical social sciences. Its use ranges from emphasizing how nature and society are co-productive of one another, to highlighting how emancipatory processes of knowledge creation necessarily involve different agents, experiences, and world views (Swyngedouw, 1999; Leff, 2015; Moragues-Faus & Marsden, 2017). In the social innovation literature, co-production is broadly defined as an active involvement of citizens and professionals in the production process of policies and knowledge (Verschuere et al., 2012). In a recent systematic review of the literature on co-production, Voorberg et al. (2015) identify three degrees of citizen involvement: citizens as co-implementers, as co-designers, or as initiators (see also Chapters 3 & 7, this volume).

Second, *reflective practices* constitute a keystone of action research processes. This literature identifies three modes of reflective practices (Kemmis & McTaggart, 2005; Bjorn & Boulus, 2011): first-person action research, which highlights the participants' ability to develop a critical approach to research practice and her/his way of being in the world; second-person research, which brings participants together to discuss shared concerns; and third-person research, where the inquiry is extended to large groups with lower levels of engagement and communication. Recently, the food studies literature has also highlighted the importance of these practices by championing reflexive governance as an approach that can open spaces for learning and adapting social solutions to collectively resolve food insecurities and sustainability challenges (Rocha, 2009; Pereira & Ruysenaar, 2012; Marsden, 2013). The central argument within this approach is that, through self- and social questioning (reflexivity), people are able to engage with the contemporary uncertainties, conflicts, inequalities, and complexities that characterize food systems (Edwards et al., 2002; Moragues-Faus et al., 2017).

In this chapter, I analyse different modes of co-production and reflective practices mobilized to conceptualize and measure progress towards sustainability in urban food systems. Ultimately, this exercise will allow us to understand how action research might contribute to addressing key challenges in the development of food assessment exercises, mainly: integrating different types of knowledge, addressing conflicting values and interests, and contributing to food system transformation.

A participatory approach to assess sustainable food systems: the case of the Sustainable Food Cities Network

The Sustainable Food Cities Network (SFCN) is a partnership project that brings together public, private, and third sector organizations that believe in the power of food as a vehicle for driving positive social change and therefore are committed to promoting sustainable food for the benefit of people and the planet. The project is coordinated by three pioneering UK NGOs: the Soil Association, Sustain, and Food Matters. In 2011, the SFCN was launched with five UK urban food policy pioneers: Bristol, Plymouth, Brighton, London, and Manchester. At the moment, the network convenes in 55 places across the UK with the goal of sharing challenges, exploring practical solutions, and developing best practices in all aspects of sustainable food. The SFCN thus constitutes a unique experience to scale up and out urban food strategies, by providing the network with the capacity to forge trans-local alliances to intervene in national food policy arenas as well as contribute to spreading good practice across UK local areas. For that purpose, it has developed a set of activities, mainly: communication platforms, networking events, campaigns, tools to share experience and training, funding schemes for members, and an awards system (Moragues-Faus, 2017a).

In order to expand this work, the SFCN was particularly interested in measuring progress towards sustainability in cities for three main reasons.

First, the SFCN promotes a holistic approach to food system reform by working across sectors and with different actors through food partnerships. While this multi-actor and cross-sectoral approach is increasingly adopted in urban food policies worldwide (Moragues-Faus & Morgan, 2015), understanding the impact of this collaborative approach on sustainability outcomes remains a shared challenge (Clark et al., 2015). Consequently, an assessment framework that shows the connections and synergies between sectors and actors can contribute to supporting more holistic and participatory interventions in the food system. Project participants were particularly interested in highlighting, through a common framework, collective efforts towards improving sustainability outcomes by reshaping urban foodscapes. Second, developing a holistic assessment framework, underpinned by shared definitions and measurements of 'progress', can help to improve and guide the development of effective urban food strategies as well as contribute to identifying gaps and inform decision-making. At the same time, an assessment framework can embrace a place-based but not place-restricted approach, by collecting experiences from different geographies and providing a flexible tool that avoids prescriptive or blanket approaches to food system reform. This is particularly important in a context of austerity that requires prioritization of actions. Finally, it is necessary to provide evidence of how holistic and collaborative approaches to urban food transformation, as well as specific food initiatives, result in sustainability and food security outcomes. This includes compiling experiences and lessons learned from different geographies. Robust evidence can also be instrumental in attracting new institutions and sources of funding to support and strengthen food partnerships.

Taken together, these motivations led to the development of the action research project analysed in this chapter. This action research process was co-designed between a Cardiff University researcher and SFCN coordinators and consisted of five steps described below.

1. *A review of both academic and 'grey' literature*: This exercise allowed us to identify the range of indicators used to measure the environmental, social, and economic sustainability of urban food strategies. This review was published on the SFCN website (Prosperi et al., 2015) and contributed to framing the participatory workshops.
2. *Participatory workshops*: We organized five workshops to define what constitutes success for an urban food strategy and to select the most relevant indicators that can guide action towards that goal. The workshops were organized geographically (Cardiff, London, Edinburgh, and two in Liverpool) in order to grasp spatial issues that might affect involvement with urban food policies at the national level (in the UK as a whole, but also Wales and Scotland) and maximize city governments' and civil society organizations' participation in the process. Participants from all the SFCN partnerships were invited to these meetings. A total of 99

stakeholders from 41 different places participated in these workshops, including policymakers (32), civil society organizations (40), private sector (6), and academics (21). The workshops consisted of the following steps:

a. Visioning: Collective definition of a sustainable food city.
b. Discussing and agreeing on key concepts: After presenting key findings from the literature review an overall framework was agreed. This included a combination of a theme-based framework and a logic framework (see Figure 6.1). The themes selected align with the broad dimensions of sustainability: economic, health and well-being, environment, and governance. The logical framework defines four levels: goals (i.e. overarching aim, ultimately what we want to achieve), outcomes (a state or position which is reached in order to achieve the goal), indicators (a measure of progress towards delivery of an outcome which shows an increase/improvement/change in movement in a relevant and measurable parameter) and activities (an action that is carried out or a step that is taken to contribute to delivering specific outcomes). Also, a set of key criteria to select the indicators was also agreed upon in each workshop using the SMART model (Specific, Measurable, Achievable, Realistic, and Time-related).
c. Defining outcomes and indicators for each dimension: Using the results from the visioning exercise around a sustainable food city, participants defined specific outcomes and indicators for each sustainability dimension.
d. Prioritization of indicators by mobilizing agreed criteria: Participants reflected on the usefulness of indicators according to the criteria previously defined.

3. *Verification of the feasibility and coherence of the selected indicators*: The results of the workshop were complemented with relevant literature, previous experience and case studies from the SFCN, and insights from meetings and interviews with local and national policymakers. After this initial feedback, SFCN coordinators and the lead researcher prepared a document for consultation that included the toolbox and that was shared widely through the SFCN (see Moragues-Faus et al., 2016). At the UK level, we received written feedback from public health officials, city councils, and civil society organizations. Furthermore, a webinar with 42 attendees and a workshop with academics and practitioners in London were conducted to gather further feedback. In total, more than 70 participants provided comments to refine the indicators toolbox.
4. *Application of the toolbox*: The set of indicators identified was applied in Bristol and Cardiff, two cities involved in the network. This step included gathering information from different institutions and databases produced in the last five years to feed the indicators and record the impact of specific policies and interventions. Data collection was carried

out in collaboration with local authorities and civil society organizations through requests of information to key informants on specific topics. The results provided a substantial and unique contribution to ongoing efforts to measure the impact of urban food strategies and contribute to refining further the indicators toolbox.

5. *Dissemination*: Dissemination has occurred throughout the project through participation in events and conferences, website communication, social media, and webinars.

The main output of this process is a toolbox for action that has two main purposes. First, to provide practitioners – and particularly local governments, policymakers, and public health decision makers – with a comprehensive collation of relevant evidence and indicators of success of a place-based approach to food. Second, to assist practitioners to plan, implement, monitor, and evaluate the impact of urban food strategies. The toolbox presents a framework that is sufficiently flexible to account for differing local circumstances and priorities. This place-based approach to assess progress in urban food systems towards sustainability is represented in Figure 6.1. Table 6.1 shows an example of how this toolbox can be implemented.

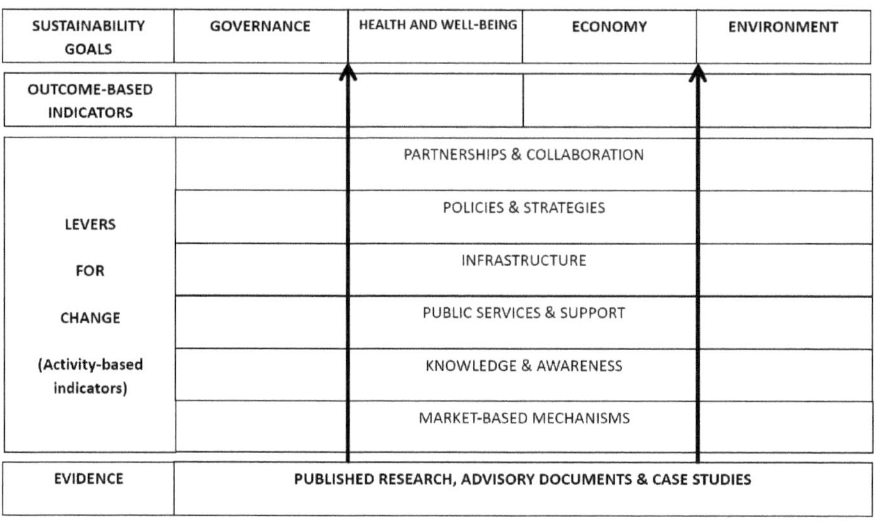

Figure 6.1 Framework to measure progress in sustainable food systems.
Source: (Moragues-Faus et al., 2016).
Note: The levers and associated activities have not been ranked in terms of importance.

Table 6.1 Examples of the application of the toolbox

Elements of the toolbox	*Examples for Health and Well-being*
Sustainability goals	Improving physical and mental health and well-being by reducing food poverty; improving access to affordable healthy food; promoting healthy weight and healthy diets; and increasing participation in food-related physical and social activity.
Outcomes (meta-indicators)	Decrease in the number of people overweight or obese.
Activities classified in 6 levers of change (Proxy indicators)	Example from lever market-based mechanisms: More healthy options are available in supermarkets, convenience stores, restaurants, takeaways, cafes, vending machines, and catering settings.
Evidence	*Example of supporting evidence from grey and scientific literature* Reference: The Food Foundation. January 2016. 'Force-Fed. Does the food system constrict healthy choices for typical British families?' The Food Foundation. The report studies the extent to which unhealthy eating patterns are becoming more common as we increasingly 'eat out'. In addition to planning regulations to limit the number of fast food outlets (mentioned above), the report recommends many other measures to 'incentivize food service providers to provide healthier food'. These include 'setting upper limits for the formulation of processed foods' for specific nutrients; making Government Buying Standards mandatory for all public procurement; using VAT to support healthy choices; increasing school meal uptake and improving food in schools and workplaces (pp. 27, 35). *Example of supporting evidence: Case Study* A shop offering healthy fast food has been opened with the help of Haringey Council (p. 7). The Council is supporting local youngsters to choose a better option by investing in a not-for-profit social enterprise that offers a tasty, healthier alternative using free-range, local, and seasonal produce.

Grounding co-productive and reflective practices

The action research process with the SFCN included co-productive and reflective practices. *Co-production* took place by involving practitioners as initiators, co-designers and co-implementers. First, the demand to conduct this research process emerged from SFCN facilitators, who were actively involved in evaluating the impact of the SFCN programme. Not only did these facilitators struggle to assess the impact of the overall programme, but individual cities also voiced the difficulty of measuring and communicating how specific food partnerships were contributing to deliver sustainability outcomes in their respective places.

Second, practitioners were *co-designers* of the research process and of the toolbox. The co-design of the research process included network facilitators from Food Matters, the Soil Association, and Sustain and also input from six food partnerships. These practitioners, together with researchers, constituted the core action research team and jointly developed the stages of the research project (described in the section, 'A participatory approach to assess sustainable food systems: the case of the Sustainable Food Cities Network'). This included the design and delivery of workshops, collection and analysis of data and feedback, participation in dissemination activities, and writing up results. Consequently, decision-making throughout the research process was collaborative. This was made possible by an early alignment of values between researchers and practitioners to guide the research process. These values are closely linked to key principles of action research and reflexive governance as discussed below (see Dick, 2002; Lenoble & De Schutter, 2010), mainly: flexibility, inclusivity and participation, horizontality, and contribution to social change as an overarching goal.

A key aspect of this research process was to develop a flexible and responsive approach to the research design as well as the outputs to confer adaptive capacity to changing organizational but also personal circumstances. Indeed, throughout the research project, personal and professional circumstances, such as a member of staff leaving the organization, led to changes in the team and reorganization of the timetable. This flexibility was underpinned by trust and open communication amongst the core team members.

Similarly, the core action research team was committed to ensuring inclusivity of different voices and needs within the SFCN membership. These participatory processes contribute to share and co-produce knowledge and information as well as build trust (IPES-Food, 2017). In this particular case, inclusion mechanisms comprised the development of participatory spaces such as workshops or webinars, involving local administrations, civil society organizations, public health practitioners, and local experts. Furthermore, the geographical spread of the workshop aimed to facilitate the participation of different actors and include their perspectives in the definition of a sustainable food city and associated measurements. This spatial and sectoral inclusivity allowed us to develop a place-based approach to food system assessment where different cities and their contrasting urban food systems could be represented under a common umbrella. That is, participants working on different food challenges – from food poverty to carbon emissions, and under diverse circumstances – big/small cities with different levels of deprivation, informed the development of a common framework that at the same time reflected their needs and priorities. This is particularly important in a context where comparison amongst cities is on the rise, through a series of national and international awards and also monitoring exercises (Moragues-Faus, 2017d). A place-based approach can contribute to the acknowledgement of the current uneven urban foodscape, where cities greatly differ in terms of average household income or levels of civil society involvement (Moragues-Faus & Sonnino, 2018). Acknowledging these different social-ecological configurations of cities, and therefore their distinct capacities and limitations to transform the urban foodscape, is essential to avoid creating a competitive landscape of winning and losing cities. In this regard, not only the inclusion of diverse voices was important but also the incorporation of case studies developed by distinct cities as part of the toolbox's evidence base supporting specific interventions in urban foodscapes.

The research team was committed to developing inclusive spaces of participation but also to assuring meaningful participation of practitioners in decision-making arenas. For that purpose, and following action research principles (Borda & Rahman, 1991), the methodology sought to develop symmetrical, horizontal, or non-exploitative relations among participants. For example, this was achieved by ensuring participants were able to shape the process as well as the conceptual framework and the indicators. As described above, each workshop provided an opportunity for participants to define sustainability from scratch, identify what type of assessment is useful for their local area by defining criteria to select indicators, and actually identifying specific outcomes and associated indicators to measure them. In this line, the overall project was considered by the research core team as a tool to promote positive social change, by empowering cities in defining progress on sustainable food systems and supporting people's collective praxis. The compilation of sustainability discourses, indicators, and also case studies from across the UK that make up the toolbox illustrate this diversity and collaboration. This overarching aim guided the development of a flexible, participative, inclusive, horizontal, place-based, and transformative approach to develop food system assessments.

Finally, co-production practices also include the *implementation* of the toolbox by practitioners. A first pilot was conducted in Bristol and Cardiff, with local experts applying the framework. However, a user-friendly version is under development by the SFCN to support wider implementation.

Reflective practices were also integrated throughout the action research process. These practices were mostly restricted to second-person action research involving the core research team made up of a Cardiff researcher and SFCN facilitators. In this regard, there were formal and informal spaces dedicated to reflecting on the process and related outputs. Activities to gather feedback at the end of every workshop and dissemination events also constituted a way to provide open spaces of deliberation and integrate third person action research within the process. For example, a planner provided the following feedback on the toolbox, 'This looks OK to me although I am disappointed our good practice publication for planners was not referenced'. On another line, a health practitioner stated, 'I think the draft document is extremely helpful and makes a lot of sense from my perspective working from within a local authority and making essential partnerships across the city to progress healthy and more sustainable food'.

Despite the benefits of this co-productive and place-based approach to food system assessment there are also key limitations and trade-offs that emerged throughout the process. First, the development of the indicators toolbox departed from the definition of sustainability, and adapted it to a particular context – UK cities. In every workshop, participants started responding to the broad question, 'what is a sustainable food city?', allowing for multiple perspectives to emerge and qualify sustainability. Responses were varied. For example, in Edinburgh participants were highly interested in governance aspects compared to other workshops. This reflects among others how the Scottish food policy context differs from UK dynamics. For instance, the Scottish government is actively engaged in addressing pressing challenges such as the right to food. Also, participants decided to focus on health and well-being as a way to encapsulate the social dimension of sustainability but also stressing the importance of health inequalities and related interventions in UK food strategies. Indeed, Public Health practitioners have been key actors shaping urban food policies in the UK in terms of content, activity, and leadership. This flexibility was instrumental in broadening participation in the definition of the framework, which assured the development of a tool that responds to people's needs and increased ownership of this resource by participants. While this exercise contributes to developing a useful toolbox that can be implemented in a specific local context, it might limit its applicability in other geographies as well as posit challenges in aligning changes at the local level with national and international dynamics and targets such as the Sustainable Development Goals. Indeed, grounding assessment frameworks on the needs and experiences of participants and potential users actively shapes those frameworks around specific challenges that might not be shared across geographies and constituencies, for example the key role that health plays in the SFCN toolbox in contrast with social dimensions of sustainability or cultural aspects.

Second, we devised a participative process with the aim to be inclusive. However, given the topic of the workshops, participation was informally restricted to those with interest and/or knowledge in measuring progress on sustainable food cities. The overall purpose of this framing was to build on the experience and knowledge of diverse frontline practitioners rather than reinventing the wheel and developing from scratch the indicators that have already been used and implemented. However, this framing does limit the representation of different knowledges and sustainability practices that are seldom recognized in standardized indicators, assessment exercises, or even in mainstream sustainability discourses (Johnston et al., 2011; Moragues-Faus, 2017b). This failure to include a wider diversity of actors might contribute to reinforce partial accounts of sustainability where more powerful or better positioned stakeholders (e.g. highly educated white middle classes) set up the agenda. Inevitably, this posits key questions around the ultimate purpose of food assessment exercises and their transformative capacity, that is, how these assessments actually reify, shift, or empower particular framings and practices around food and sustainability.

Finally, this project, and particularly the application of the toolbox in Cardiff and Bristol, showed that it is extremely challenging to develop urban food-related indicators that are accessible, reliable, comprehensive, and inexpensive to collect. There are important data gaps in outcome-related indicators, particularly to characterize local food economies and the environmental impact of local food systems. Consequently, the process of selecting indicators includes navigating complex trade-offs. In this context, we sought to develop a useful, flexible toolbox that recognizes its limitations. In our case, some of the indicators are more accessible than others. For example, it is easy to monitor the decrease in the number of overweight or obese people, and more difficult to measure changes in the number of jobs in the local food economy. However, by including in the framework activities that can be implemented to deliver sustainability outcomes we aim to provide proxies that contribute to fill these information gaps. Nevertheless, the framework does not establish a direct causality link between the implementation of specific activities (i.e. healthier school meals) and specific outcomes (i.e. reduction on childhood obesity). This is due to the multiple causes involved in producing these outcomes (such as obesity) and the myriad processes shaping local foodscapes at the same time (e.g. increase of unhealthy take-aways around schools). Finally, this framework would also benefit from breaking down indicators in terms of neighbourhoods, gender, age, income, and ethnic background to account for the diversity within cities.

Conclusions: place-based, reflexive, and co-productive practices as a tool for social change

This chapter has analysed the action research process that led to the development of an indicators toolbox for sustainable food cities as a means to overcome key challenges in current food assessment exercises, these being mainly: inclusivity of different knowledges, addressing conflicting values and interests, and supporting food system transformation. In this regard, the combination of a systems approach to food sustainability and a place-based perspective is a key contribution of this framework. Indeed, by collaboratively building a flexible framework, the toolbox succeeds in providing a non-prescriptive tool that accommodates the diversity of UK cities and recognizes that there are various pathways towards sustainability. In this line, the toolbox allows one to understand different contexts and progressive change by providing not only a fixed picture but also depicting the evolution of particular urban food systems. For that purpose, outcome-based indicators depict progress (increase/decrease) of particular trends, from malnutrition to the importance of local food business. This place awareness recognizes (and celebrates) differences in initial resources and capabilities of each place. The inclusion of voices from different sectors and geographies provides a flexible tool that encourages the ability to innovate and adapt to changing circumstances. This diversity is also reflected in the evidence base that supports the overall framework, which includes academic and grey literature but also specific case studies from across the UK. Furthermore, the framework not only assesses progress, but its implementation helps map current activities, connect different sectors and stakeholders in the city, and inspire action.

Despite the multiple contributions of this framework to address current food system assessment challenges, there are limitations that emerged throughout the process. These revolve around including visions and needs of a wider (and mostly underrepresented) set of stakeholders, unpacking the different power relations at play that condition the framing of food system assessments, and providing efficient and reliable measures that can be easily applied and communicated at different scales. While all knowledge is provisional and necessarily contextualized, it is also relevant to map the connections between different assessment exercises, as a means to reinforce a trans-local food movement based on sharing and co-producing knowledge (Blay-Palmer et al., 2016). The exercise discussed in this chapter takes a loose definition of local and urban, allowing food partnerships implementing the tool to define the boundaries. It also exemplifies how food assessments can contribute to scaling up and out good practices, by offering a tool to co-develop across local experiences within the same country and at the same time codifying knowledge to replicate and inspire action. This cross-scalar capacity and flexibility provides advantages but also requires further exploration of the interdependencies between places and scales. These interdependencies call for a further problematization of the role played by proximate but also distant foodscapes in configuring more or less sustainable places. Taken together, these limitations call for further experimentation of critical, participatory and reflective practices in the definition of frameworks and measurements.

Overall, the need to embed more critical, co-productive, and reflexive approaches within food assessment exercises mimics a wider call for implementing reflexive and participative governance principles in the food system to deliver food security and sustainability outcomes. Other chapters in this volume reflecting on a myriad of case studies also stress this need to rethink governance processes in the development of sustainable food system assessments (see also Chapters 2, 3, 5, 8, 10, & 11, this volume). Reflexive governance entails addressing conflicting values and power imbalances as well as their implications in the food system (Moragues-Faus et al., 2017). But also, it requires an explicit examination of the genesis of these conflicting interests as well as the role that each actor plays in reproducing or hindering food system failures. It is therefore necessary to create the conditions for such processes to unfold, as Lenoble & De Schutter (2010) state:

> This (...) requires not simply an environment that is empowering and facilitative, but also a transformation in the understanding of the actors themselves of how they should redefine their roles (...) This reflexivity may be understood as the 'internalisation' of the conditions of learning. Learning, in other terms, should be conceived as an operation in which the actors themselves redefine their understanding of the problem to be addressed, and are led not simply to question the solutions that are routinely explored, but also their relationship to the problem and the way they traditionally define their interests.
>
> (p. 30)

The development of these critical, reflexive, and co-productive practices needs to expand from conceptual debates to permeate governance arenas and food system assessment exercises, but also research processes. After all, knowledge production and assessment frameworks constitute a powerful tool to frame debates and current food system challenges and potential solutions that constitute an integral part of the governance of food systems (Mooney & Hunt, 2009; Nally, 2014; Moragues-Faus, 2017c). In this context, we academics have the opportunity to surpass and enrich conceptual debates by committing to developing deliberative and collaborative research processes as a means to actively transform our food system.

Acknowledgements

I would like to thank all the individuals and groups actively working to transform the UK food system, especially those who took time to participate in this action research process. Special thanks to Alizee Marceau, Tom Andrews, and Claire Devereux for their active involvement in this project. Also I would like to acknowledge the financial support of the Economic and Social Research Council (ESRC) in the UK, the Welsh Government and the European Commission through the Ser Cymru II Research programme. Finally, I am very grateful to the editors of this volume for their constructive comments. All responsibility for the content of the chapter remains with the author.

Notes

1 To access the toolkit and associated indicators see http://www.fao.org/in-action/food-for-cities-programme/toolkit/introduction/en/.

2 The Milan Food Policy Pact Indicators Framework can be accessed here: http://www.milanurbanfoodpolicypact.org/milan-urban-food-policy-pact-monitoring-framework/.

References

Alrøe, H.F., Moller, H., Læssøe, J., & Noe, E. (2016) Opportunities and challenges for multicriteria assessment of food system sustainability. *Ecology and Society*. 21(1), p. 38. Available at: http://dx.doi.org/10.5751/ES-08394-210138.

Alrøe, H.F., Sautier, M., Legun, K., Whitehead, J., Noe, E., Moller, H., & Manhire, J. (2017) Performance versus values in sustainability transformation of food systems. *Sustainability*. 9(3), p. 332. Available at: https://doi.org/10.3390/su9030332.

Bjorn, P. & Boulus, N. (2011) Dissenting in reflective conversations: Critical components of doing action research. *Action Research*. 9(3), pp. 282–302. Available at: https://doi.org/10.1177/1476750310396949.

Blay-Palmer, A., Sonnino, R., & Custot, J. (2016) A food politics of the possible? Growing sustainable food systems through networks of knowledge. *Agriculture and Human Values*. 33(1), pp. 27–43.

Borda, O.F. & Rahman, M. (1991) *Action and knowledge: Breaking the monopoly with participatory action-research*. New York, Apex Press.

Carlsson, L., Callaghan, E. , Morley, A., & Broman, G. (2017) Food system sustainability across scales: A proposed local-to-global approach to community planning and assessment. *Sustainability*. 9(6), 1061. Available at: https://doi.org/10.3390/su9061061.

Clark, J.K., Sharp, J.S., & Dugan, K.L. (2015) The agrifood system policy agenda and research domain. *Journal of Rural Studies*. 42, pp. 112–122.

Conijn, J.G., Bindraban, P.S., Schröder, J.J., & Jongschaap, R.E.E. (2018) Can our global food system meet food demand within planetary boundaries? *Agriculture, Ecosystems & Environment*. 251, pp. 244–256.

Dick, B. (2002) Postgraduate programs using action research. *The Learning Organization*. 9 (4), pp. 159–170. Available at: https://doi.org/10.1108/09696470210428886.

Dubbeling, M., Santini, G., Renting, H., Taguchi, M., Lançon, L., Zuluaga, J., de Paoli, L., Rodriguez, A., & Andino V. (2017) Assessing and planning sustainable city region food systems: Insights from two Latin American cities. *Sustainability*. 9(8), 1455. doi:10.3390/su9081455.

Edwards, R. , Ranson, S., & Strain, M. (2002) Reflexivity: Towards a theory of lifelong learning. *International Journal of Lifelong Education*. 21(6), pp. 525–536.

FAO (2014) *SAFA sustainability assessment of food and agriculture systems-guidelines version 3.0*. Rome, Food and Agriculture Organization (FAO).

Godfray, H.C.J., Crute, I.R., Haddad, L., Lawrence, D., Muir, J.F., Nisbett, N., Pretty, J., Robinson, S., Toulmin, C., & Whiteley, R. (2010) The future of the global food system. *Philosophical Transactions of the Royal Society B: Biological Sciences*. 365(1554), pp. 2769–2777.

Gustafson, D., Gutman, A., Leet, W., Drewnowski, A., Fanzo, J., & Ingram, J. (2016) Seven food system metrics of sustainable nutrition security. *Sustainability.* 8(3), 196. Available at: https://doi.org/10.3390/su8030196.

Hanson, L.L. & Schrader, D. (2014) Creating new urban spaces of sustainability and governmentality: An assessment of the development of a food and urban agriculture strategy for Edmonton, Canada. In *From sustainable to resilient cities: Global concerns and urban efforts* (pp. 191–214). Bingley, UK, Emerald Group Publishing Limited.

Hatt, S., Artu, S., Brédart, D., Lassois, L., Francis, F., Haubruge, É., Garré, S., Stassart, P.M., Dufrêne, M., Monty, A., & Boeraeve, F. (2016) Pour des systèmes agricoles et alimentaires durables: Le concept de l'agroécologie et comment il questionne les pratiques actuelles de recherche (synthèse bibliographique). *Biotechnologie, Agronomie, Société et Environnement* 20(S1), pp. 215–224.

Hiremath, R.B., Balachandra, P., Kumar, B., Bansode, S.S., & Murali, J. (2013) Indicator-based urban sustainability – a review. *Energy for Sustainable Development.* 17 (6), pp. 555–563.

Ilieva, R.T. (2017) Urban food systems strategies: A promising tool for implementing the SDGs in practice. *Sustainability.* 9(10), 1707. Available at: https://doi.org/10.3390/su9101707.

Ingram, J. (2011) A food systems approach to researching food security and its interactions with global environmental change. *Food Security.* 3(4), pp. 417–431.

IPES-Food (2017) *What makes urban food policy happen? Insights from five cities.* International Panel of Experts on Sustainable Food Systems. Available at: http://www.ipes-food.org/_img/upload/files/Cities_execsummary.pdf. Accessed 10/04/2019.

Johnston, J., Szabo, M., & Rodney, A. (2011) Good food, good people: Understanding the cultural repertoire of ethical eating. *Journal of Consumer Culture.* 11(3), pp. 293–318.

Johnston, J.L., Fanzo, J.C., & Cogill, B. (2014) Understanding sustainable diets: A descriptive analysis of the determinants and processes that influence diets and their impact on health, food security, and environmental sustainability. *Advances in Nutrition.* 5(4), pp. 418–429.

Kemmis, S. & McTaggart, R. (2005) *Communicative action and the public sphere.* Thousand Oaks, CA, Sage.

Landert, J., Schader, C., Moschitz, H., & Stolze, M. (2017) A holistic sustainability assessment method for urban food system governance. *Sustainability.* 9(4), 490. Available at: https://doi.org/10.3390/su9040490.

Leff, E. (2015) The power-full distribution of knowledge in political ecology. In: Perreault, T., Bridge G., & McCarthy, J. (Eds.) *The Routledge Handbook of Political Ecology.* London & New York, Routledge, pp. 64–75.

Lenoble, J. & De Schutter, O. (2010) *Reflexive governance: Redefining the public interest in a pluralistic world.* Oxford, Hart Publishing.

Lynch, A., Andreason, S., Eisenman, T., Robinson, J., & Steif, B. (2011) *Sustainable urban development indicators for the United States.* University Park, PN, Penn University.

Mahon, N., Crute, I., Simmons, E., Islam, M.M. (2017) Sustainable intensification – 'oxymoron' or 'third-way'? A systematic review. *Ecological Indicators.* 74, pp. 73–97.

Marsden, T. (2013) From post-productionism to reflexive governance: Contested transitions in securing more sustainable food futures. *Journal of Rural Studies*. 29(0), pp. 123–134.

Mooney, P.H. & Hunt, S.A. (2009) Food security: The elaboration of contested claims to a consensus frame. *Rural Sociology*. 74(4), pp. 469–497.

Moragues-Faus, A. (2017a) Addressing food security vulnerabilities through urban food policy alliances: The case of the sustainable food cities network in the UK. In: Barling, D. (Ed.) *Advances in food security and sustainability*. Amsterdam, Elsevier, pp. 87–113.

Moragues-Faus, A. (2017b) Emancipatory or neoliberal food politics? Exploring the 'politics of collectivity' of buying groups in the search for egalitarian food democracies. *Antipode*. 49(2), pp. 455–476. Available at: http://doi.wiley.com/10.1111/anti.12274.

Moragues-Faus, A. (2017c) Problematising justice definitions in public food security debates: Towards global and participative food justices. *Geoforum*. 84, pp. 95–106.

Moragues-Faus, A. (2017d) Report meeting of city food networks MUFPP2017. Milan Urban Food Policy Pact Annual Gathering and Mayors' Summit, Valencia, 21 October. Available at: https://drive.google.com/file/d/1i1Y0zlxTEX04GzLreGUjImdDTDJ3DMJZ/view. Accessed 10/04/2019.

Moragues-Faus, A. & Marsden, T. (2017) The political ecology of food: Carving 'spaces of possibility' in a new research agenda. *Journal of Rural Studies*. 55, pp. 275–288.

Moragues-Faus, A. & Morgan, K. (2015) Reframing the foodscape: The emergent world of urban food policy. *Environment and Planning A*. 47(7), pp. 1558–1573.

Moragues-Faus, A. & Sonnino, R. (2018) Re-assembling sustainable food cities: An exploration of translocal governance and its multiple agencies. *Urban Studies*. 56(4), pp. 778–794.

Moragues-Faus, A. , Marceau, A., & Andrews, T. (2016) Making the case and measuring progress: Towards a systems approach to healthy and sustainable food. Report of the ESRC financed Project 'Enhancing the Impact of Sustainable Urban Food Strategies'. Available at: http://sustainablefoodcities.org/Portals/4/Documents/SFC%20indicators%20final%20draft%20for%20website.pdf. Accessed 10/04/2019.

Moragues-Faus, A., Morgan, K., Moschitz, H., Neimane, I., Nilsson, H., Pinto, M., Rohracher, H., Ruiz, R., Thuswald, M., Tisenkopfs, T., & Halliday, J. (2013) Urban food strategies: The rough guide to sustainable food systems. Report available at: http://www.foodlinkscommunity.net/fileadmin/documents_organicresearch/foodlinks/publications/Urban_food_strategies.pdf.

Moragues-Faus, A., Sonnino, R., & Marsden, T. (2017) Exploring European food system vulnerabilities: Towards integrated food security governance. *Environmental Science and Policy*. 75, pp. 184–215.

Nally, D. (2014) Governing precarious lives: Land grabs, geopolitics, and 'food security'. *The Geographical Journal*. Available at: http://doi.wiley.com/10.1111/geoj.12063.

OECD (Organisation for Economic Cooperation and Development) (2015) *In it together: Why less inequality benefits all*. Paris, OECD. Report available at: http://www.oecd-ilibrary.org/employment/in-it-together-why-less-inequality-benefits-all_9789264235120-en.

Pereira, L. & Ruysenaar, S. (2012) Moving from traditional government to new adaptive governance: The changing face of food security responses in South Africa. *Food Security*. 4: 41–58. Available at: https://doi.org/10.1007/s12571-012-0164-5.

Prosperi, P. , Moragues-Faus, A. , Sonnino, R., & Devereux, C. (2015) *Measuring progress towards sustainable food cities: Sustainability and food security indicators*. Report of the ESRC financed Project 'Enhancing the Impact of Sustainable Urban Food Strategies'. Available at: http://sustainablefoodcities.org/Portals/4/Documents/Measuring%20progress%20towards%20sustainable%20food%20cities_final%20report%20w%20appendixes.pdf.

Reason, P. & Bradbury, H. (2008) Inquiry & participation in search of a world worthy of human aspiration. In: Reason, P. & Bradbury, H. (Eds.) *Handbook of action research: participative inquiry and practice*. London, SAGE, pp. 1–64.

Rocha, C. (2009) Developments in national policies for food and nutrition security in Brazil. *Development Policy Review*. 27(1), pp. 51–66. Available at: http://doi.wiley.com/10.1111/j.1467-7679.2009.00435.x.

RUAF (2016) Developing tools for mapping and assessing sustainable city region food systems (CityFoodTools). Available at: https://www.ruaf.org/projects/developing-tools-mapping-and-assessing-sustainable-city-region-food-systems-cityfoodtools. Accessed 10/04/2019.

Schader, C. , Grenz, J. , Meier, M.S., & Stolze, M. (2014) Scope and precision of sustainability assessment approaches to food systems. *Ecology and Society*. 19(3), 42. Available at: http://dx.doi.org/10.5751/ES-06866-190342.

Seekell, D., Carr, J., Dell'Angelo, J., D'Odorico, P., Fader, M., Gephart, J., Kummu, M., Magliocca, N., Porkka, M., Puma, M., Ratajczak, Z., Rulli, M.C., Suweis, S., & Tavoni, A. (2017) Resilience in the global food system. *Environmental Research Letters*. 12(2). Available at: http://doi.org/10.1088/1748–9326/aa5730/.

Sustainable Development Solutions Network (2015) Indicators and a monitoring framework for the sustainable development goals launching a data revolution for the SDGs. Available at: www.jstor.org/stable/resrep15877.

Swyngedouw, E. (1999) Modernity and hybridity: Nature, regeneracionismo, and the production of the Spanish waterscape, 1890–1930. *Annals of the Association of American Geographers*. 89(3), pp. 443–465.

Tanguay, G., Rajaonson, J., Lefebvre, J., & Indicators, P.L.-E. (2010) Measuring the sustainability of cities: An analysis of the use of local indicators. *Ecological Indicators*. 10(2), pp. 407–418.

Travers, K.D. (1997) Nutrition education for social change: Critical perspective. *Journal of Nutrition Education*. 29(2), pp. 57–62.

Underwood, T., McCullum-Gomez, C., Harmon, A., & Roberts, S. (2011) Organic agriculture supports biodiversity and sustainable food production. *Journal of Hunger & Environmental Nutrition*. 6(4), pp. 398–423.

Verschuere, B., Brandsen, T., & Pestoff, V. (2012) Co-production: The state of the art in research and the future agenda. *VOLUNTAS: International Journal of Voluntary and Nonprofit Organizations*. 23(4), pp. 1083–1101.

Voorberg, W.H., Bekkers, V.J.J.M., & Tummers, L.G. (2015) A systematic review of co-creation and co-production: Embarking on the social innovation journey. *Public Management Review*. 17(9), pp. 1333–1357.

7 Building consensus on sustainable food system assessment

Applying a Delphi survey

Paolo Prosperi, Thomas Allen, and Bruce Cogill

Introduction

Food security and sustainability are at the forefront of political agendas, increasingly stressing the need to improve our understanding and capacity to leverage the linkages between food, health, and the environment. The scientific community has been strongly encouraged by policymakers and various stakeholders to assess the multiple impacts of the food we produce, process, consume, and waste.

Food system approaches specifically aim to step up to this challenge, promoting interdisciplinary and multi-stakeholder analyses of the nexus between diets, the environment, and human health (see also Chapters 10 & 11, this volume). Although there is a host of valuable efforts that identify tools for the assessment of the sustainability of food – both at the national and the international levels, there remains a lack of consensus around metrics to measure sustainable food systems (Perignon et al., 2017). An agreed information system is key for evidence-based knowledge to guide and assess actions. While this need for a limited set of universal indicators may be clear, the method and selection of specific indicators is not. This chapter reports the process used to develop a framework and select a suite of technically and conceptually sound indicators.

In 2013, Bioversity International and the International Center for Advanced Mediterranean Agronomic Studies-Mediterranean Agronomic Institute of Montpellier (CIHEAM-IAMM) with support from the Daniel and Nina Carasso Foundation, have established a multidisciplinary taskforce of experts to identify a shortlist of indicators of 'Sustainable Diets and Food Systems'. Building on past efforts and expertise, in particular on the need to integrate nutritional and agro-biodiversity challenges to the sustainability debate (Fanzo et al., 2012; FAO & Bioversity International, 2012; Johnston et al., 2014), both institutions have worked to identify a framework and select suitable indicators involved in assessing sustainable food systems.

Building on an existing vulnerability and resilience framework, the team conducted a large expert consultation, through two focus groups and a

Delphi survey, which allowed a systematic and reproducible identification of indicators for the assessment of sustainable food systems. A Delphi survey is an iterative social science technique for opinion gathering, recognized as an appropriate approach to build consensus and prioritize indicators from different academic fields.

The chapter first discusses the concept of sustainable food systems and presents a conceptual framework that can be adapted to articulate key factors and outcomes of food systems. It then presents the process and synthesizes the main results of the expert consultation, and finally discusses the key lessons learned in designing an interdisciplinary research programme and identifies sustainable food systems metrics adapted to a specific geographical context.

Sustainable food systems: a multidisciplinary concept

The multidimensional nature of sustainable development

Sustainable development is multidimensional; it has to satisfy several economic development, social equity, and environmental protection goals. The most frequently quoted definition of sustainability comes from 'Our Common Future', also known as the 'Brundtland Report' (UN, 1987). Human development must meet 'the needs of the present without compromising the ability of future generations to meet their own needs'. When applied to the agricultural and food sector, Conway's (1986) frequently quoted definition of agro-ecosystem sustainability refers to 'the ability of a system to maintain productivity in spite of a major disturbance, such as caused by intensive stress or a large perturbation' (p. 35). As a property of a system, sustainability becomes open to interactions with the external. In other words, sustainability is the dynamic preservation, over time, of the intrinsic identity of the system among perpetual changes (Gallopín, 2003).

Interdisciplinary efforts implicating life science, earth and environmental science, agriculture and nutrition, and social and sustainability science[1] require a better understanding of the interactions of global change and food security, and not merely integrating multiple bio-physical and socio-economic factors into the analysis (see also Chapter 2, this volume). There is a need to further cross and link current evidence and knowledge across various disciplines. For instance, in the current context of rapid change, measures of food and nutrition security that only focus on outcomes – such as hunger and malnourishment – are too narrow to capture the dynamics of transformation within food systems.

Multiple factors influence the course of human–environment interactions, which are further complicated by the presence of coevolving causal forces. Research in both the natural and social sciences uses the idea of a system to explain complex dynamics (see also Chapter 4, this volume). A system is a network of multiple elements that are interconnected through causal relationships. Modern societies depend on complex systems to provide food. Food systems encompass an array of cyclical activities from soil through to waste recycling and disposal management – including production, processing, packaging and distribution, retailing and consumption – and involve a multitude of actors (Ericksen et al., 2009). Such complexity calls for approaches engaging experts, academics, and multiple stakeholders. For this reason, at every phase of our project, an expert consultation was carried out and the multidisciplinary composition of the expert panel was considered a key element of reliability of the process and constantly monitored.

An integrated set of indicators

Assessing the sustainability of complex systems thus implies taking into account a wide range of dimensions and indicators. Scholars and policymakers have been calling for the development of an integrated system of food security and sustainability to inform decision-making (Barrett, 2010; Dicks et al., 2013). Sustainability metrics must encompass a wide array of issues relevant to human existence and nature, and they must be useful in guiding the system towards a sustainable trajectory.

Feenstra et al. (2005) define an indicator as 'something used to show the condition of a system' and Gallopín (1997) notes it as an 'operational representation of an attribute (quality, characteristic, property) of a system'. Indicators must, however, be well grounded in science and allow for comparisons across different systems. There is a need for a comprehensive and evidence-based suite of indicators to lead public policy interventions and provide information on adaptive management that is necessary for the practical implementation of sustainability. Indicators, or metrics, gathered as an organized information system and dynamically combined to provide a perspective, target three principal objectives:

- Inform civil society, industry, public officials, and all stakeholders
- Measure impact or progress towards defined goals
- Aid decision-making processes

A sound theoretical framework is the starting point in constructing metrics (OECD, 2008). The selection of sustainability indicators is generally realized following the guidelines of a conceptual framework and a series of criteria related to the availability and quality of the data. However, the selection should be based on what is desirable to measure and not only on which indicators are available. Concurrently, the exercise of developing indicators should also take into account limitations, such as budget constraints, that can make measurement and replication difficult over time. Furthermore, ethics, the transparency, as well as the reproducibility, of the whole exercise is essential in constructing credible indicators.

Framing workable hypotheses

Background conceptual framework

The High Level Panel of Experts on Food and Nutrition Security (HLPE) provided a succinct definition of a sustainable food system by interconnecting the previously coined concepts of food security (UN, 1996) and sustainable development (UN, 1987): 'A sustainable food system (SFS) is a food system that delivers food security and nutrition for all in such a way that the economic, social, and environmental bases to generate food security and nutrition for future generations are not compromised' (HLPE, 2014). Under this definition, the specification of the generational-temporal factor emphasizes the link between food systems and their ability to maintain or enhance their functions over time (Prosperi et al., 2014).

Developing a multidimensional conceptual framework to explore the sustainability of food systems implies specifying what is meant by sustainability. The research approach in this chapter builds on the assumptions that sustainability assessment aims at capturing the ability of a system to maintain and enhance its essential functions over time (i.e. feeding people properly), and that sustainability addresses threats to preserving life support systems, including their capacity to environmentally, socially, and economically withstand and adjust (Allen & Prosperi, 2016). Since Ericksen's 2008 article, 'Conceptualizing food systems for global environmental change research', it has been often agreed that several global and regional biophysical and socio-economic drivers of change affect the structure and processes of food systems, and thus contribute to, or put at risk, context-specific food and nutrition security outcomes. Fulfilling food system outcomes remains challenging because of socio-economic and biophysical stressors affecting the food system. Food systems are then considered social-ecological systems – with economics strongly included in the social dimension – that comprise biophysical and social factors linked through feedback mechanisms (Allen et al., 2014a).

Food systems as social-ecological systems

Foran et al. (2014) comprehensively synthetize what is generally intended by the term 'social-ecological system' (SES). They note, 'SES visualizes the human-environment interface as a coupled "system" in which socio-economic as well as biophysical driving forces interact to influence food system (and sub-system) activities and outcomes, both of which subsequently influence the driving forces' (p. 90). Thus, SESs are complex and dynamic systems that are continuously adapting in response to internal or external pressures. They involve societal, human, economic, and ecological subsystems in mutual interactions (Schlüter et al., 2014).

This understanding of food systems as social-ecological systems helps answer questions about the sustainability problems that affect the functions of the food system. Food and nutrition security is, in fact, considered the principal outcome of any food system and relies on several properties and activities of food systems, namely: producing, processing and packaging, distributing, retailing, and consuming. Those food systems' activities and properties emanate, in turn, in three main sets of outcomes such as food and nutrition security (availability, access, utilization), social welfare, and environmental capital.

Furthermore, various elements of food systems are altered by, and actively impact, the socio-economic and environmental conditions of the system across local, regional, and global levels. Food security depends on multiple bio-physical drivers – such as land and water resources degradation, biodiversity loss, sea-level rising – and economic and social stresses – such as demographic dynamics, technological innovation, economic trends, and social changes – that interact with each other and then impact, individually or concurrently, different aspects of the food system. Such socioeconomic and biophysical drivers might impact food security directly or indirectly, positively or negatively, and change over time; when a food system fails to deliver people food security, considered as its primary outcome, the system can be considered as vulnerable.

Therefore, food systems can be vulnerable, and resilient, to a set of stressors (Adger, 2006) such as environmental pressures, socio-economic instabilities and institutional and policy factors (Turner et al., 2003). The identification of vulnerability and resilience variables can help to proxy the metrics of food systems. These variables will be determinant in explaining if the system is able to meet over time its foremost outcome, that is, ensuring food and nutrition security for all. They will point to the social, economic, and environmental elements of the system which condition, and can be leveraged to guarantee, the availability of healthy and accessible food for human nutrition.

Vulnerability and resilience as properties of food systems

Sustainability addresses threats to preserving life support systems, including their capacity to withstand and adjust. It is then crucial to assess stocks of, and changes in, human and natural assets. Derived from sustainability sciences (Turner et al., 2003), the vulnerability and resilience approach within the social-ecological systems frameworks, proved relevant to analyse the sustainability of critical food and nutrition security outcomes (Prosperi et al., 2014; Allen & Prosperi, 2016; Prosperi et al., 2016).

Vulnerability is the degree to which a system is likely to experience harm due to exposure to a perturbation or stress. Resilience represents the ability of a system to anticipate, absorb, accommodate, or recover from the effects of a potentially hazardous event, in a timely and efficient manner, through ensuring the preservation, restoration, or improvement of its essential basic structures and functions. Since Turner et al.'s 2003 'A framework for vulnerability analysis in sustainability science', (2003), a strong effort has been made to encourage the consideration of vulnerability and resilience research in sustainability science. Theories of vulnerability and resilience proved helpful to understand the complex dynamics involving socioeconomic and biophysical aspects, and to implement sustainable development strategies and research programmes.

Building on the vulnerability and resilience theoretical framework, a causal-factor approach allows studying the sequential pathway through which food systems' key outcomes can be threatened. Exposure, sensitivity, and resilience[2] become key food system properties defining its capacity to ensure food and nutrition security over time (Figure 7.1). Understanding the causal mechanisms that regulate the interactions between drivers of change and food and nutrition security issues can help in analysing and interpreting available information, developing metrics, and anticipating new hazards and changes. The investigation of causes, effects, and responses to socio-economic and biophysical changes can provide analytical tools to further understand the problems that affect the sustainability of the food system. The conceptual hierarchical framework developed has been operationalized for modelling the complex relationships between food and nutrition security and sustainability in a specific geographical context, developing indicators of sustainable food systems (Allen et al., 2014b). To

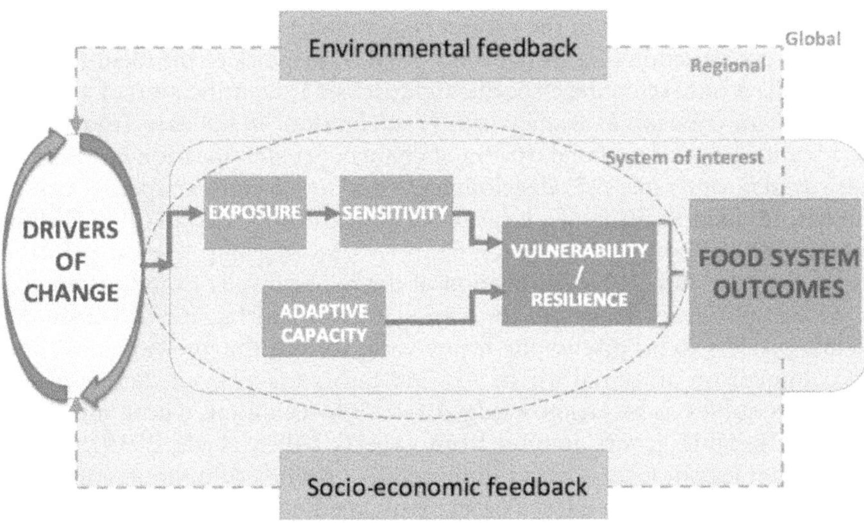

Figure 7.1 A sustainable food system framework.
Source: Adapted from: Allen and Prosperi, 2016.

illustrate the application of the framework, the approach was applied to countries in the Mediterranean region.

Discussing and selecting indicators

Focus groups and the Delphi study: an expert-based approach

Sustainability can have different understandings and participatory approaches might be particularly appropriate to identify and share the conditions, priorities, and resource constraints crucial to sustainability assessment. Participatory processes with stakeholder involvement are often acknowledged to be crucial for sustainability assessment. Implication of key actors and disciplines should be embedded in all steps, leading to a co-production of knowledge from problem definition towards local solutions (see also Chapters 2, 3, 5, & 8, this volume). Concurrently, food system practitioners call for appropriate metrics from researchers and the identification of indicators for policymaking needs to be traceable, evidence based, and scientifically sound in order to guarantee transparency in decision-making and effectiveness of evidence-based policy (Bell & Morse, 2013). Qualitative consultation methods, such as focus group techniques and Delphi surveys, allow for a robust participatory process and satisfactorily shared results. Specifically, our project's main outcome consists of gathering and synthesizing scientific knowledge for the assessment of sustainable food systems, thus it involves mainly scientists and academics (from several disciplines) in order to provide practitioners with suitable metrics.

Building on multidisciplinary[3] and interactive research practices, the methodology developed in the project is composed of ten steps (Figure 7.2), from the construction of a global conceptual framework to the identification of a reduced suite of context-specific indicators: (1) identification of a global food system conceptual framework; (2) definition of the case study area; (3) identification of essential drivers of change; (4) identification of essential food system outcomes; (5) development of a set of context-specific causal models; (6) identification of a large set of indicators; (7) design of the Delphi survey; (8) elicitation of feedback from two focus-group sessions; (9) the Delphi survey; and (10) identification of the reduced suite of indicators.

Two exploratory focus groups were gathered and facilitated before the Delphi process to (a) discuss the framework, (b) test the questionnaire, and (c) comment on an initial list of 136 indicators taken from the literature. A Delphi survey is an acknowledged research technique whose aim is to obtain a reliable group opinion from experts (Allen et al., 2019). It is a group interaction process directed in iterative rounds of opinion collection and feedback and provides a systematic method to involve experts in problem analysis and discussion on complex issues, helping convert diverse views and opinions into one or more communal notions through an iterative feedback process (Allen et al., 2019). The survey was conducted via email

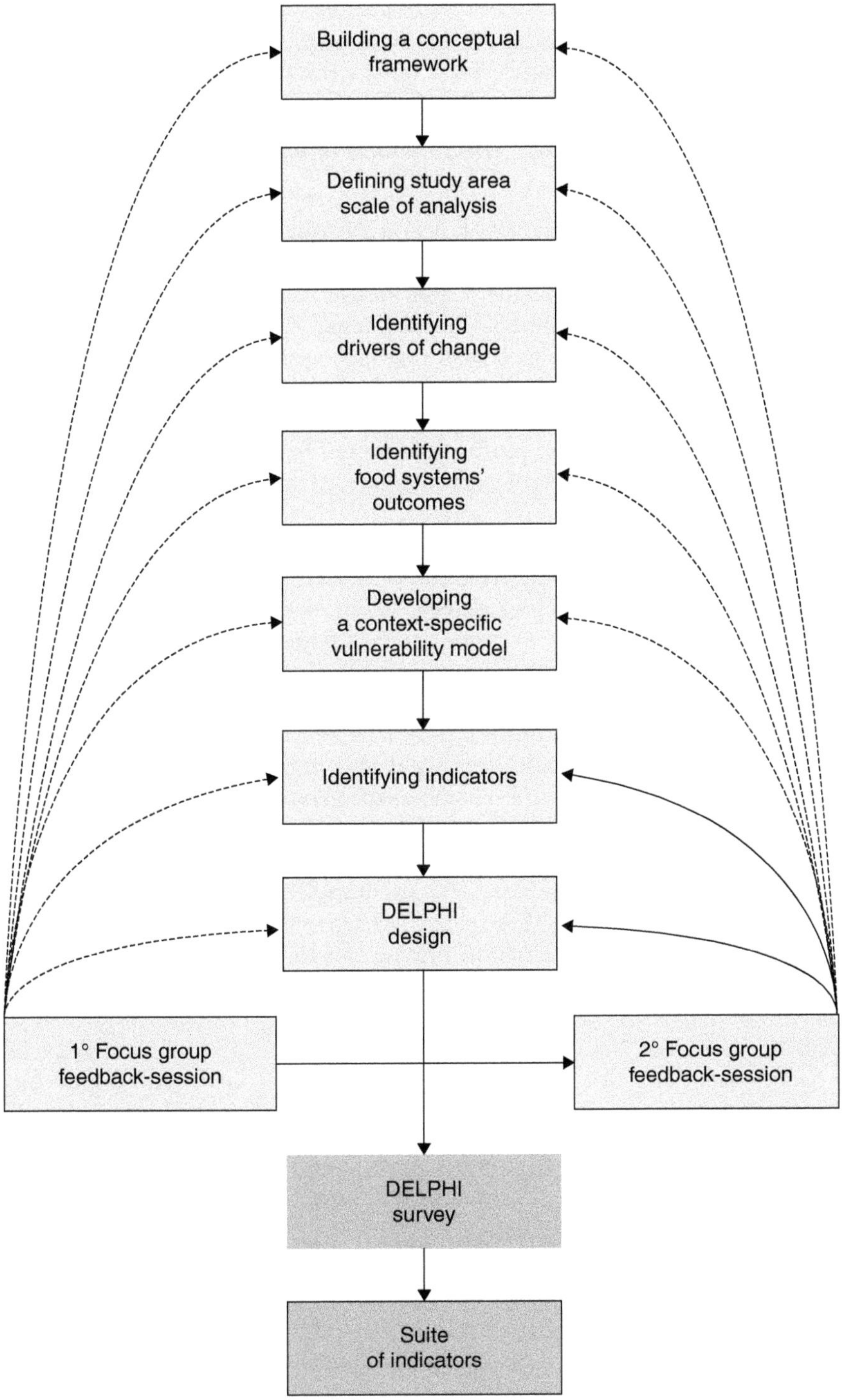

Figure 7.2 A sequence of 10 steps.

Source: Authors' elaboration modified from Allen et al., 2019.

and SurveyMonkey®, a web-based survey platform, with the aim of further discussing the survey outcomes in an ex-post workshop with all participants.

Identification of eight selected causal models of vulnerability and resilience

Following an extensive review of literature, the vulnerability/resilience-adapted framework for the global food system (Figure 7.1) was proposed as the basis for discussion during a first focus group with a panel of multi-disciplinary experts. The vulnerability/resilience framework was proposed as suitable for our research as it was regularly quoted by multiple sources discussing food systems sustainability and presented a broad hierarchical system of information that could serve as a starting point for discussion. The identification of a causal pathway (adapted from Metzger & Schroeter, 2006) allowed locating the role of the three variables of exposure, sensitivity, and resilience.

Following the first focus group, eight specific causal models of vulnerability and resilience were selected within a larger set of models. Shaping the interactions where a set of drivers of change – that is, *water depletion; biodiversity loss; food price volatility; changes in food consumption patterns* – directly affect a set of food and nutrition security outcomes at a sub-regional level – that is, *nutritional quality of food supply; affordability of food; dietary energy balance; satisfaction of cultural food preferences.* Those drivers of change, as well as food and nutrition security outcomes and the related interactions, are specific to the geographical area of the Latin Arc within the Mediterranean region (for a justification of the geographical scale and information on local food system characteristics see Allen and Prosperi, 2016). Each interaction was disentangled in exposure, sensitivity, and resilience. In particular, these sets of characteristics are indicating how changes in water, biodiversity, food prices, and food consumption patterns are transmitted through the regional food system. This includes the sequence of events and the scale of interactions: (a) how the regional food system is sensitive to these changes; and (b) the subsequent adaptive capacity of the food system (see Box 7.1 for brief definitions of these main issues and drivers; see Figure 7.3 and Table 7.1 for both a graphic and an analytical description of the eight causal dynamics).

Box 7.1 Proposed drivers and issues

Drivers

Water depletion is 'a use or removal of water from a water basin that renders it unavailable for further use' (Molden, 1997). Water availability is closely related to climate change trends altering precipitation patterns and rainwater (SCAR, 2011). It is also related to agrofood patterns and the use and concentration of agrochemicals, impacting the quality of water, further contributing to water scarcity.

Biodiversity loss is defined as 'the long-term or permanent qualitative or quantitative reduction in components of biodiversity and their potential to provide goods and services, to be measured at global, regional and national levels' (Convention on Biological Diversity, 2004). Biodiversity loss is cogenerated by climate change, environmental depletion, and water stress. It is strongly related to modern food production and consumption patterns (Altieri, 2000) that have become more intensive and homogenizing.

Food price volatility refers to large and atypical 'variations in agricultural prices over time' (FAO, 2011). Climate change, changing trade patterns, new dietary trends, and growing demand for biofuels are often invoked as causes of food price volatility. Speculation on commodity markets and reduction of food stocks are also crucial determinants of price variations (Robles et al., 2009).

Changes in food consumption patterns refer to the changing structure of global food consumption, related to changing dominant values, attitudes, and behaviours (Kearney, 2010; Johnston et al., 2014). Individual food consumption patterns – that is, diets – are the results of changes in culture, social values, and representations attached to food consumption, driving effectively behavioural changes and resulting in modified diets. The global changes in food consumption patterns – with a shift to more animal-sourced products, and foods high in fat, energy, and salt (Drewnowski and Popkin, 1997) – are largely driven by demographic factors and income growth, and are related to changes in activity levels, lifestyle, globalization, urbanization, markets, changes in occupational status and employment distribution, and more effective dissemination of information (Meade, 2012).

Food and nutrition security issues

Nutritional quality of the food supply refers to the nutritional composition of the food products on the market (Observatoire de la qualité de l'alimentation – Oquali, Institut national de la recherche agronomique – INRA). The improvement of the nutritional quality of the food supply is one of the eight specific actions defined by the Word Health Organization European Action Plan for Food and Nutrition Policy 2007–2012. A balanced diet is achieved through personal habits but also requires that the foods eaten by consumers have a satisfactory nutritional composition.

Affordability of food is defined as 'the purchasing power of households or communities relative to the price of food' (Ingram, 2011). It refers to the 'economic access' to food (Foran et al., 2014). Affordability is about food being available at prices that people can afford to pay, and in particular, whether low-income consumers can afford to buy enough nutritious and healthy food to meet basic needs (Barling et al., 2010).

Dietary energy balance refers to the balance between energy intake and energy expenditure (Patel et al., 2004). Excessive fat accumulation is acknowledged to be a risk factor for various health problems, including cardio-vascular diseases, diabetes, cancers, and osteoarthritis (WHO, 2008). A range of environmental, social, and behavioural factors interact to determine energy intake and expenditure, such as sedentary lifestyles, consumption of and heavy marketing of both energy-dense foods and fast food outlets, adverse social and economic conditions, the consumption of high-sugar drinks, etc. (Swinburn et al., 2004).

Cultural food preferences are powerful environmental factors related to social background and behaviours, which contribute to food choices and intakes. Recognizing ethnic and cultural food preferences and changes, compatible with nutritional requirements, is essential for food acceptance and well-being. Food preferences, socially or culturally determined, are now recognized as a key consideration in food security.

Next, a large list of indicators was identified by the research team and discussed in a second focus group with the same expert panel. Additional indicators were proposed by the experts, while some were deleted, and the process resulted in a shortlist of 136 indicators. Both the suggested framework and the shortlist of indicators were then submitted to a large panel of experts for further discussion though a Delphi consultation.

Identification of indicators

Following the protocol of the Delphi survey (Allen et al., 2019), 52 experts from more than 40 academic and policy institutions worldwide were asked to discuss and refine the framework and the underlying assumptions, and to test the framework by selecting proxy indicators.

First, an extensive list of 213 potential experts was developed by reviewing academic publications. An electronic letter of invitation was sent to the identified experts to explain the goals and protocol of the study, and permitted potential participants to self-estimate their expertise and aptness to the study. Two weeks later, a general email was sent to all identified experts containing a link to the questionnaire and background material.

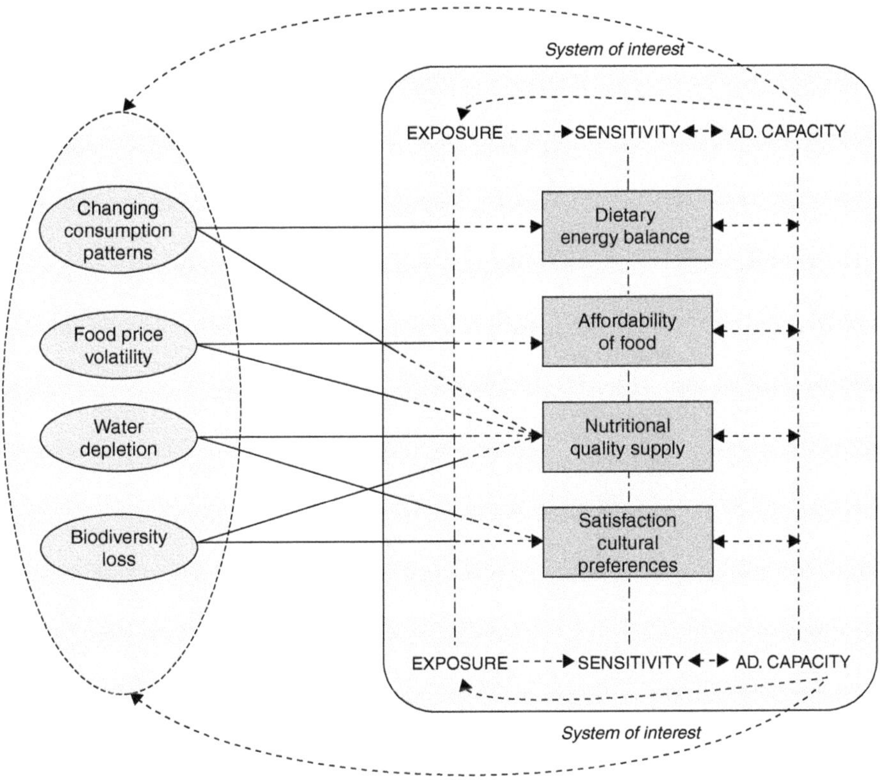

Figure 7.3 Interacting drivers and outcomes – graphic description.
Source: Allen and Prosperi, 2016.

This material included a document explaining the conceptual background, the specific aim and purpose of the Delphi study, and the summary details of the proposed indicators.

After each round a statistical analysis was run to provide participants with feedback to revise the questionnaire. Feedback reports providing each participant with the group results and their individual previous responses were sent via email after each of the three Delphi rounds. Overall, the final results were presented four months after sending the first letter of invitation. In each round, participants were asked to select their preferred indicator for each of the 24 components of the framework from a menu of five to eight preselected indicators (see Table 7.2). Twenty-four indicators are the desired outcomes from the selection of three indicators (i.e. exposure, sensitivity, resilience) per interaction analysed (i.e. eight selected interactions between drivers of change and food security issues). Participants had the opportunity to propose new indicators. Indicators that did not receive any

Table 7.1 Interacting drivers and outcomes – analytical description

Drivers of change	*Food and Nutrition Security Issues*		*References*
	Nutritional quality of food supply		
WATER DEPLETION	Potential Impact	• Contributing to decreasing production and productivity of nutritious foods. • Engendering low dilution capacity and contamination of agri-food products. • Impacting the availability of quality foods for poor consumers through higher cost of water.	SCAR, 2011; Wood et al., 2010; PARME, 2011.
	Recovery Potential	• Fostering water productivity to guarantee adequate nutritional values of foods. • Contrasting water scarcity through agrobiodiversity richness. • Enhancing adaptation through food import from water rich countries. • Reusing wastewater safely for use as water sources. • Focusing on human capacities and institutional framework.	SCAR, 2011; UNWATER, 2014.
	Affordability of food		
	Potential Impact	• Altering productivity, prices, and trade, and then food availability and affordability. • Increasing water prices lead to higher costs of agrofood production and to decrease in food affordability.	Wood et al., 2010; SCAR, 2011.
	Recovery Potential	• Encouraging drought-resistant crop utilization. • Fostering food import from water rich countries. • Improving irrigation efficiency. • Promoting waste water treatments.	Hellegers et al., 2008; Waughray, 2011.

	Nutritional quality of food supply		
BIODIVERSITY LOSS	Potential Impact	• Shifting to ecologically simplified systems based on cereals, which contributes to poorly diversified diets. • Hampering food systems responses against climate change, with consequent impact on productivity. • Increasing the dependency on global varieties on external inputs.	Arimond et al., 2010; SCAR, 2011.
	Recovery Potential	• Promoting agrobiodiverse systems for ecosystem services, food security benefits (nutritional value of foods), the viability of agricultural systems, and long-term productivity. • Fostering organic farming.	Thrupp, 2000; Reidsma & Ewert, 2008.
	Satisfaction of cultural food preferences		
	Potential Impact	• Putting at risk cultural traditions and preferences, linked to regional varieties and diets. • Homogenizing food production. • Contributing to reduce the enormous amount of information on nutritional and health benefits of the foods that shape the food cultural preferences of people. • Decreasing food biodiversity, which could result in the loss of unique and traditional foods.	Kearney, 2010; SCAR, 2011.
	Recovery Potential	• Knowing how to prepare a more varied diet can influence consumption of different food products. • Providing more varied and tasteful diets. • Enhancing and keeping traditional food cultures.	Termote et al., 2010; Johnston et al., 2014.
FOOD PRICE VOLATILITY	Potential Impact	• Impacting food production and consumption. • Altering food supply towards disadvantaged groups. • Leading to profound changes in the composition and availability of food supplies. • Hampering the present agrofood system supply, strongly interlinked with the fossil fuels system.	DEFRA, 2008; SCAR, 2011.

(*continued*)

Table 7.1 (Cont.)

Drivers of change	*Food and Nutrition Security Issues*		*References*
	Recovery Potential	• Enhancing dietary diversity for avoiding dependency on few groups of foods. • Fostering local provisioning and production, less involved in price variations.	Pinstrup-Andersen, 2013.
	Affordability of food		
	Potential Impact	• Impacting household incomes and purchasing power. • Affecting agrofood productivity, and therefore food affordability and availability. • Exacerbating economic shocks for the poor, who depend on wages and the rest of the economy. • Shifting purchasing strategies to lower quality products.	Wood et al., 2010; HLPE, 2011; SCAR, 2011; Regmi & Meade, 2014.
	Recovery Potential	• Fostering food industry's focus on consumers and their need for 'affordable food of high quality and diversity'. • Shifting towards cheaper or locally available foods, meeting the same caloric and nutritional requirements. • Implementing food policies for diversifying supply sources through different strategies (subsidies, food stamps). • Promoting diversity in food consumption patterns.	European Technology Platform, 2008; Brunori & Guarino, 2010.
	Nutritional quality of food supply		
CHANGES IN FOOD CONSUMPTION PATTERNS	Potential Impact	• Influencing food industry production patterns, overall food security, and nutritional characteristics of diets. • Shifting the demand towards cereals, simple sugars, animal products, and highly processed foods.	European Technology Platform, 2008; Brunori & Guarino, 2010; SCAR, 2011; UNEP, 2012.
	Recovery Potential	• Improving the understanding of the determinants of consumer choices. • Empowering consumers' choice for healthy and safe provided food. • Engendering consumption patterns cognizant of the impact of food choice on health.	SCAR, 2011; Allen et al., 2014.

<u>**Dietary Energy Balance**</u>		
Potential Impact	• Increasing consumption of fats, sugars, sweeteners, animal products, highly processed foods, and in fast foods and vending machines products. • Decreasing consumption in plant proteins and of home-prepared foods. • Strengthening 'obesogenic' environments with little energy expenditure and sedentary lifestyles. • Altering frequency and the amounts consumed of foods. • Decreasing dietary diversity.	PARME, 2011; SCAR, 2011; UNEP, 2012.
Recovery Potential	• Fostering public awareness for healthier diets through campaigns and community movements. • Enhancing cultural knowledge on preparing varied diets and on nutritional and health benefits of the foods. • Promoting weight loss and metabolic health through appropriate changes in the gut microbiota. • Supporting guidelines on dietary strategies to counteract overweight and obesity.	Barling et al., 2010; Termote et al., 2010; Lopez-Legarrea et al., 2014.

participant preference were excluded from the following rounds. New indicators were added if at least two participants proposed the same, or a similar, variable. A 'Don't know' option was always included in the menu to allow experts to express their lack of knowledge on a specific component.

After a first open-ended round, panellists were presented with the opportunity to justify or amend their first choices. Succeeding rounds have been designed to bring the group to focus or consensus. An upgraded framework and a restricted set of indicators were reached, after three rounds, from this consultation process. The Delphi study revealed low-, medium-, and high-consensus and a majority-level on indicators in 75 per cent of the interactions out of the 24 initial ones. The results obtained in terms of global response, expert participation rates, and consensus on indicators, were then satisfactory. Also, experts confirmed with positive feedback the appraisal of the components of the framework.

Consensus was finally reached in round three for 14 of the 24 desired indicators (see Table 7.2). Eight indicators have met the high threshold consensus criteria (80 per cent), three other indicators have met the medium threshold consensus criteria (70 per cent), and another three have achieved the low threshold consensus criteria (60 per cent). Four indicators have been selected by the majority of the participants (above 50 per cent). For five dimensions (out of 24), clear bi-dimensionality can be reported (two indicators above 35 per cent). In some of these cases, several experts recommended constructing a composite indicator. Three dimensions remained unresolved with a wide dispersion of expert opinions among indicators and little improvement of the consensus through the rounds (see final round results in Table 7.2).

The chosen list of 14 indicators includes:

1. Water Footprint of nutrient-dense foods [cubic metres/kg]
2. Intensity of use of actual water resources [%]
3. Irrigation Water Efficiency Index [%]
4. Water Footprint for an average diet [cubic meters/yr]
5. % of total acreage of top 5 varieties
6. Nutritional Functional Diversity
7. Crop Agrobiodiversity Factor
8. % of diets locally produced
9. % of nutrient intakes (Vit. A, Zn, I, Fe) from 10 most volatile foods
10. Household Dietary Diversity Score
11. % of food household expenditure
12. Sensitivity to price volatility
13. Food Purchasing Power Index
14. Household Dietary Diversity Score
    Prevalence of overweight and obesity is just below minimum threshold consensus criteria (60 per cent).

Table 7.2 Level of consensus reached by indicator

		Nutritional Quality of Food Supply	*Affordability of Food*
Water Depletion	Exposure	75% Water Footprint of nutrient-dense foods	86% Water Footprint for an average diet
	Sensitivity	61% Intensity of use of actual water resources	53% Price index for 10 most water-demanding foods
	Resilience	83% Irrigation Water Efficiency Index	47% Cross-price elasticity of demand of high/low water demanding foods
		Nutritional Quality of Food Supply	**Satisfaction of Cultural Food Preferences**
Biodiversity Loss	Exposure	64%	47%
		% of total acreage of top 5 varieties	Import Dependency Ratio
	Sensitivity	83%	72%
		Nutritional Functional Diversity	% of diets locally produced
	Resilience	89%	53%
		Crop Agrobiodiversity Factor	Integration of biodiversity considerations in business
		Nutritional Quality of Food Supply	**Affordability of Food**
Price Volatility	Exposure	72%	81%
		% of nutrient intakes from 10 most volatile foods	% of food household expenditure
	Sensitivity	47%	86%
		Price elasticity of 10 most nutrient-dense foods	Sensitivity to price volatility
	Resilience	92%	53%
		Household Dietary Diversity Score	Presence of safety net programmes
		Nutritional Quality of Food Supply	**Dietary Energy Balance**
Change in Food Consumption Patterns	Exposure	64%	47%
		Food Purchasing Power Index	Caloric share of ready-to-consume products
	Sensitivity	83%	58%
		Household Dietary Diversity Score	Prevalence of overweight and obesity
	Resilience	28%	28%
		% of public expenditure on food subsidies AND Existence of national dietary guidelines	Existence of policy plan for overweight/obesity AND Funding allocated to nutrition education

Lessons learned

Discussing implementation of the framework

This study aims at filling the theoretical and methodological gaps in quantitative assessment of sustainable food systems, combining a theory-driven approach with expert judgment, rather than a data-driven process. Since 'what is badly defined is likely to be badly measured' (OECD, 2008), efforts were concentrated on the operationalization of theories (vulnerability and resilience) in order to build a solid, common and replicable basis for defining a metric system. Within a broad systemic approach, the research attempted to operationalize the framework for the assessment of the sustainability in food systems through multidisciplinary and multi-stakeholder consultation. This research effort is provided to the scientific community, practitioners, or policymakers who might be interested in assessing and disentangling the characteristics of a given food system through the operationalization of this dynamic framework.

First, one challenge was to identify pathways leading to vulnerability, and the characteristics and opportunities ensuring resilience of the food system in a context of change. Resilience and vulnerability are considered problematic to operationalize through precise assessment methods due to their theoretical and multidimensional nature. It emerged from the workshop convened after the Delphi survey that participants had sometimes an incomplete understanding of the proposed framework. This is a shortcoming of the operationalization of the vulnerability and resilience framework as it has been already observed by working with practitioners (Foran et al., 2014) and it can have an impact on the indicator selection.

Second, several participants would have liked to have seen other food systems' outcomes than food and nutrition security issues considered. As it was already emerging from the Delphi consultation, environmental and social outcomes are standing out as crucial elements to consider and include in the assessment exercise. It was highlighted that this would be more in line with the general perception of what sustainability means: 'People think about sustainability as an outcome'. 'People want a descriptor of a state rather than the prediction of a state'. Furthermore, some experts would also have liked to complement the list of food and nutrition issues, adding elements such as 'dietary quality'.

The use of the food system framework developed was nevertheless acknowledged to anticipate and predict possible future outcomes of the food systems. A participant presented the framework as 'a model', highlighting the causal pathway that it aimed at providing. Some participants recalled that 'understanding what is driving the outcomes is important'.

Informing policy towards sustainable food systems

Assessing issues related to sustainability problems, with the goal of informing the decision-making process, has a number of critical implications. There is, in fact, a growing debate about the importance of the role, utility, adoption, focus, and final goals of the sustainability indicators. There are several different ways to interpret indicators and select data. It is therefore important to know how the information provided by the indicators is going to be transferred to policymakers, and what the actual aims are of using the indicators. Aggregation of data can strongly alter the messages for policymakers, and several studies demonstrate that often the indicators that have been prepared in an appropriate technical manner are not actually applied nor do they have a real impact on policymaking (Bell & Morse, 2013; see also Chapters 5 & 11, this volume).

Referring to the European Union institutions, Sébastien and Bauler (2013) proved also the need for a greater involvement of the actors of the political and institutional contexts where indicators have to be identified and applied. A strong and active involvement of the local/community stakeholders is key to designing a set of metrics that will be useful to measure real progress and gaps towards the sustainability of food systems (see also Chapter 6, this volume). Moreover, the theories of vulnerability and resilience are often acknowledged as particularly effective by the scientific community for both conceptual and methodological aims of research, while development practitioners find those theories difficult to operationalize, with local actors at a context-specific level, for their complex and systemic nature (Foran et al., 2014). However, practitioners consider metrics as crucial tools to measure development and sustainability goals achieved in a given food system (Dicks et al., 2013).

Another important question is the type of policymakers targeted and the role of the media in informing policymakers. 'Who are the stakeholders we need to influence?' 'Who are the policy makers?' An expert suggested that there may be different goals for policymakers at different levels, for example, (1) 'to communicate to high-level policy makers and media about the overall state of the food systems by focusing on food system outcomes', and that (2) '[i]mplementing diagnostic models and causal analyses can help food-focused policy-makers as well as other types of policy-makers'.

Conducting a Delphi survey

A number of lessons can be drawn in terms of practice to enhance validity, replicability, participation, and consensus for further Delphi studies:

- It is important to demonstrate to participants the benefits for society and science of the proposed survey, while considering the potential shortcomings of this approach
- Given the diversity of views and understandings of what sustainability means, discussions need to be guided through structured and replicable methods, in particular if metric systems are the final outcome of the discussions. In this regard, iterative approaches are appropriate

- Transparency and multidisciplinary participation are crucial in the development of sustainability indicators, but present the risk of weaker consensus. Therefore, allowing a longer time-frame for decision-making may involve trade-offs
- Agreeing on a detailed background framework is essential for the development of indicators, but unlikely to happen if the framework is either too specific or not adaptable. The objectives of informing local stakeholders and aiding decision-making should be the driving principles when reducing the framework to its core elements, prioritizing short-term decision efficiency over long-term sustainability
- Having institutional support could help participants feel the beneficial purpose for society instead of fostering an exclusively profit-seeking aim for the team running the study
- If possible, holding a face-to-face meeting would help to dissipate remaining uncertainties and possible misunderstandings. For instance during the Delphi survey it was mentioned that a technical workshop (actually held in Montpellier, France, on November 2014) would have been convened at the end of the Delphi study and that participants would have been invited for further scientific discussion and involvement
- For selecting appropriate sustainable food system metrics, it is crucial to convene a diverse and appropriate expert team with a very good knowledge and understanding of the problems of the sustainability of the food systems
- Gathering two preliminary focus-group sessions as a pilot application for a Delphi helps to conceive properly the first questionnaire, managing, motivating, and administrating feedbacks
- Sending qualitative personalized feedback with comments, explanations, and suggestions from the experts enables real interaction of the group
- The use of Internet technology allows for the opportunity to consult large, geographically dispersed, expert communities
- Providing relevant but not overloaded scientific content and materials to participants allows them to be informed participants
- Structuring the survey makes each round progressively less time-consuming
- Further efforts are needed to build context-specific vulnerability and resilience frameworks that are adaptable and suitable to effectively identify metrics with both researchers and development practitioners

Conclusion

The 'Metrics of Sustainable Diets and Food Systems Project', led jointly by Bioversity International and CIHEAM-IAMM, has contributed to the exploration of assessment approaches to develop information systems for sustainable food systems. The broad vulnerability and resilience framework has been proposed to capture the food system as a whole and identify key system elements that policy can control or mitigate. Food systems are networks in which components are connected to each other through causal pathways operating at different geographical or time scales. Distinguishing three components – exposure, sensitivity, and resilience – allows the model to specify which attributes need to be measured and how to structure the different indicators in a coherent assessment framework. The operationalization of this framework in a limited geographical area (i.e. the Latin Arc in the Mediterranean region) allowed for modelling dynamic interactions specific to the analysed region.

An innovative participatory research methodology – a Delphi survey – has implemented discussion of this framework, guiding the identification of indicators. It provided the systematic and scientific approach to propose a first core set of indicators to assess the sustainability of diets and food systems. The Delphi method, with the participation of several experts coming from different disciplines and institutions, provided practitioners, and eventually policymakers, with a transparent view of the process of developing sustainability metrics for food systems. The participation of experts was included all along the theoretical and operational research process. Before the Delphi process, two focus group feedback sessions with experts have contributed to improving the theoretical framework and tailoring the questionnaires. Expert opinion was crucial from the beginning to select the most urgent food system's drivers of change, and food and nutrition security issues, as well as to validate the dynamic interactions proposed in the framework. Also, with particular regard to the questionnaire, focus group experts helped finding the best way to address Delphi participants with questions on the set of metrics, in order to make the questionnaire more understandable and, therefore, to make the iterative process successful in terms of response rate.

This exercise has shown what is required to construct a shared information system for the assessment of sustainable food systems, replicable at different scales: (a) developing a sound and general conceptual framework of food systems outcomes and drivers, based on theories and evidence-based observations at both the global and local scale; (b) facilitating the involvement of experts in knowledge production to provide critical feedback and create consensus; and (c) identifying context-specific metrics and guaranteeing a traceable and reproducible selection process.

The process resulted in consensus on 14 indicators. Moving forward, an enhanced understanding of the availability of data to compute these indicators and of the interpretation of their results is needed.

Acknowledgements

This research process was mainly realized in 2013–2014 within the framework of a scientific collaboration between the CIHEAM-IAMM (Montpellier, France) and Bioversity International (Montpellier and Rome): Paolo Prosperi was affiliated to CIHEAM-IAMM, Thomas Allen and Bruce Cogill to

Bioversity International. Thomas Allen's work was supported by the Daniel and Nina Carasso Foundation (Grant Number: 00030240) and CGIAR Research Programme *Agriculture for Nutrition and Health*. The Daniel and Nina Carasso Foundation had no role in the design, analysis, or writing of this chapter. The authors would like to thank the 52 experts for their participation and input in the Delphi survey.

Notes

1 Sustainability science was introduced in 2001 by Kates et al. (2001).

2 Exposure is the nature and degree to which a system is likely to be affected by the occurrence of a change. Sensitivity is the degree to which a system is affected either adversely or beneficially, by a change. Resilience is the ability of a system to anticipate, absorb, accommodate, or recover from the effects of a potentially hazardous event in a timely and efficient manner, through ensuring the preservation, restoration, or improvement of its essential basic structures and functions (IPCC, 2012).

3 This project implies a multidisciplinary approach since it involves participation of academic experts from different disciplines. However it does not imply – at least at this stage – a transdisciplinary approach, since local practitioners or stakeholders are not directly involved in the participatory research process for indicators, but the effort consists of providing practice actors with metrics obtained through traceable and scientific knowledge-synthesis methods. Nevertheless, the research process was built on literature taking Mediterranean policy reports as main references.

References

Adger, W.N. (2006) Vulnerability. *Global Environmental Change*. 16(3), pp. 268–281.

Allen, T. & Prosperi, P. (2016) Modeling sustainable food systems. *Environmental Management*. 57(5), pp. 956–975.

Allen, T., Prosperi, P., Cogill, B., & Flichman, G. (2014a) Agricultural biodiversity, social-ecological systems and sustainable diets. *Proceedings of the Nutrition Society*. 73(4), pp. 498–508.

Allen, T., Prosperi, P., & Cogill, B. (2014b) *Metrics of sustainable diets and food systems*. Workshop Report. Montpellier, France, Bioversity International & CIHEAM-IAMM.

Allen, T., Prosperi, P., Cogill, B., Padilla, M., & Peri, I. (2019) A Delphi approach to develop sustainable food system metrics. *Social Indicators Research*. 141, pp. 1307–1339.

Altieri, M.A. (2000) Multifunctional dimensions of ecologically-based agriculture in Latin America. *The International Journal of Sustainable Development & World Ecology*. 7(1), pp. 62–75.

Arimond, M., Wiesmann, D., Becquey, E., Carriquiry, A., Daniels, M. C., Deitchler, M., Fanou-Fogny, N., Joseph, M.L., Kennedy, G., Martin-Prevel, Y., & Torheim, L.E. (2010) Simple food group diversity indicators predict micronutrient adequacy of women's diets in 5 diverse, resource-poor settings. *The Journal of Nutrition*. 140(11), pp. 2059S–2069S.

Barling, D., Lang, T., & Sharpe, R. (2010) The re-emergence of UK National Food Security on the policy agenda: Sustainability challenges and the politics of food supply. In: Lawrence, G. et al. (Eds.) *Food security, nutrition and sustainability: New challenges, future options*. London, Earthscan, pp. 61–78.

Barrett, C.B. (2010) Measuring food insecurity. *Science*. 327(5967), pp. 825–828.

Bell, S. & Morse, S. (2013) Towards an understanding of how policy making groups use indicators. *Ecological Indicators*. 35, pp. 13–23.

Brunori, G. and Guarino, A. (2010) 'Security for whom? Changing discourses on food in Europe in times of a global food crisis', in Lawrence, G. et al (Eds.), *Food security, nutrition and sustainability*, Earthscan, London, pp. 41–60.

Convention on Biological Diversity (2004) Conference of the parties to the Convention on Biological Diversity. Decision adopted by the Conference of the Parties to the Convention on Biological Diversity – Seventh meeting Kuala Lumpur, 9–20 and 27 February 2004.

Conway, G. R. (1986) *Agroecosystem analysis for research and development*. Bangkok, Winrock International.

Department for Environment Food and Rural Affairs (2009) *Indicators for a Sustainable Food System*, DEFRA, London, England.

Dicks, L., Bardgett, R., Bell, J., Benton, T., Booth, A., Bouwman, J., Brown, C., Bruce, A., Burgess, P.J., Butler, S.J., Crute, I., Dixon, F., Drummond, C., Freckleton, R.P., Gill, M., Graham, A., Hails, R.S., Hallett, J., Hart, B., Hillier, G.J., Holland, J.M., Huxley, J.N., Ingram, J.S.I., King, V., MacMillan, T., McGonigle, D.F., McQuaid, C., Nevard, T., Norman, S., Norris, K., Pazderka, C., Poonaji, I., Quinn, C.H., Ramsden, S.J., Sinclair, D., Siriwardena, G.M., Vickery, J.A., Whitmore, A.P., Wolmer, W., & Sutherland, W.J. (2013) What do we need to know to enhance the environmental sustainability of agricultural production? A prioritisation of knowledge needs for the UK food system. *Sustainability*. 5(7).

Drewnowski, A. & Popkin, B.M. (1997) The nutrition transition: New trends in the global diet. *Nutrition Reviews*. 55(2), pp. 31–43.

Ericksen, P. (2008) Conceptualizing food systems for global environmental change research. *Global Environmental Change*. 18(1), pp. 234–245.

Ericksen, P., Ingram, J., & Liverman, D.M. (2009) Food security and global environmental change: Emerging challenges. *Environmental Science & Policy*. 12(4), pp. 373–377.

European Technology Platform (2008) *European Technology Platform on Food for Life: Implementation action plan*. Confederation of the food and drink industries of the EU, Brussels, Belgium.

Fanzo, J., Cogill, B., & Mattei, F. (2012) *Metrics of sustainable diets and food systems, technical brief-Madrid roundtable*. Rome, Bioversity International.

FAO (2011) *The State of Food Insecurity in the World 2011: How does international price volatility affect domestic economies and food security?* FAO, Rome.

FAO & Bioversity International (2012) *Proceedings of the international scientific symposium: Biodiversity and sustainable diets united against hunger*. Rome, FAO.

Feenstra, G., Jaramillo, C., McGrath, S., & Grunnell, A.N. (2005) *Proposed indicators for sustainable food systems*. Portland, OR, Ecotrust.

Foran, T., Butler, J.R., Williams, L.J., Wanjura, W.J., Hall, A., Carter, L., & Carberry, P.S. (2014) Taking complexity in food systems seriously: An interdisciplinary analysis. *World Development*. 61, pp. 85–101.

Gallopín, G.C. (1997) Indicators and their use: Information for decision-making. Part one – Introduction. In Moldan, B. & Bilharz, S. *Sustainability indicators. A report on the project on indicators of sustainable development*. Chichester, Wiley.

Gallopín, G.C. (2003) *A systems approach to sustainability and sustainable development*. New York, United Nations Publications.

Hellegers, P., Zilberman, D., Steduto, P. and McCornick, P. (2008) 'Interactions between water, energy, food and environment: evolving perspectives and policy issues', *Water Policy*, Vol. 10, No. S1, pp. S1–S10.

HLPE. (2011) *Price volatility and food security: a report by the High Level Panel of Experts on Food Security and Nutrition of the Committee on World Food Security*, FAO, Rome, Italy.

HLPE (2014) *Food losses and waste in the context of sustainable food systems*. A report by the High Level Panel of Experts on Food Security and Nutrition of the Committee on World Food Security, FAO, Rome, Italy.

Ingram, J. (2011). A food systems approach to researching food security and its interactions with global environmental change. *Food Security*. 3(4), pp. 417–431.

IPCC (2012) *Managing the risks of extreme events and disasters to advance climate change adaptation: Special report of the intergovernmental panel on climate change*. Cambridge, Cambridge University Press.

Johnston, J.L., Fanzo, J.C., & Cogill, B. (2014) Understanding sustainable diets: A descriptive analysis of the determinants and processes that influence diets and their impact on health, food security, and environmental sustainability. *Advances in Nutrition: An International Review Journal*. 5(4), pp. 418–429.

Kates R.W., Clark W.C., Corell, R., Hall, J.M., Jaeger, C.C., Lowe, I., McCarthy, J.J., Schellnhuber, H.J., Bolin, B., Dickson, N.M., Faucheux, S., Gallopin, G.C., Grübler, A., Huntley, B., Jäger, J., Jodha, N.S., Kasperson, R.E., Mabogunje, A., Matson, P., Mooney, H., Moore, B., O' Riordan, T., & Svedin, U. (2001) Sustainability science. *Science*. 292(5517), pp. 641–642.

Kearney, J. (2010) Food consumption trends and drivers. *Philosophical Transactions of the Royal Society B: Biological Sciences*. 365(1554), pp. 2793–2807.

Lopez-Legarrea, P., Fuller, N., Zulet, M., Martinez, J. and Caterson, I. (2014) 'The influence of Mediterranean, carbohydrate and high protein diets on gut microbiota composition in the treatment of obesity and associated inflammatory state', *Asian Pacific Journal of Clinical Nutrition*, Vol. 23, No. 3, pp. 360–368.

Meade, B. (2012) *Global food markets*. Washington, DC, USDA.

Metzger, M. & Schröter, D. (2006) Towards a spatially explicit and quantitative vulnerability assessment of environmental change in Europe. *Regional Environmental Change*. 6(4), pp. 201–216.

Molden, D. (1997) *Accounting for water use and productivity*. Stockholm, IWMI.

OECD (2008) *Handbook on constructing composite indicators: Methodology and user guide*, Paris, OECD.

PARME. (2011) *Quelles recherches et quels partenariats pour la Méditerranée? Atelier de Réflexion Prospective. Rapport final*, Agropolis International, Montpellier, France.

Patel, A.C., Nunez, N.P., Perkins, S.N., Barrett, J.C., & Hursting, S.D. (2004) Effects of energy balance on cancer in genetically altered mice. *The Journal of Nutrition*. 134(12), pp. 3394S–3398S.

Perignon, M., Vieux, F., Soler, L.G., Masset, G., & Darmon, N. (2017) Improving diet sustainability through evolution of food choices: Review of epidemiological studies on the environmental impact of diets. *Nutrition Reviews*. 75(1), pp. 2–17.

Pinstrup-Andersen, P. (2013) 'Nutrition-sensitive food systems: from rhetoric to action', *The Lancet*, Vol. 382, No. 9890, pp. 375–376.

Prosperi, P., Allen, T., Cogill, B., Padilla, M., & Peri, I. (2016) Towards metrics of sustainable food systems: A review of the resilience and vulnerability literature. *Environment Systems and Decisions*. 36(1), pp. 3–19.

Prosperi, P., Allen, T., Padilla, M., Peri, I., & Cogill, B. (2014) Sustainability and food & nutrition security: A vulnerability assessment framework for the Mediterranean region. *SAGE Open*, 4(2). Available at: http://sgo.sagepub.com/content/4/2/2158244014539169.full.pdf.

Regmi, A. and Meade, B. (2013) 'Demand side drivers of global food security', *Global Food Security*, Vol. 2, No. 3, pp.166–171.

Reidsma, P. and Ewert, F. (2008) 'Regional farm diversity can reduce vulnerability of food production to climate change', *Ecology and Society*, 13(1).

Robles, M., Torero, M., & Von Braun, J. (2009) *When speculation matters*. International Food Policy Research Institute. Issue Brief 57, Washington, DC, IFPRI.

SCAR. (2011) *3rd SCAR foresight exercise: Sustainable food consumption and production in a resource-constrained world*, Standing Committee on Agricultural Research, Foresight Expert Group, European Commission, Brussels.

Schlüter, M., Hinkel, J., Bots, P.W., & Arlinghaus, R. (2014) Application of the SES framework for model-based analysis of the dynamics of social-ecological systems. *Ecology and Society*. 19(1). Available at: http://www.ecologyandsociety.org/vol19/iss1/art36/.

Sébastien, L. & Bauler, T. (2013) Use and influence of composite indicators for sustainable development at the EU-level. *Ecological Indicators*, 35, pp. 3–12.

Swinburn, B.A., Caterson, I., Seidell, J.C., & James, W.P. (2004) Diet, nutrition and the prevention of excess weight gain and obesity. *Public Health Nutrition*. 7(1a), pp. 123–146.

Termote, C., Van Damme, P. and Dhed'a Djailo, B. (2010) 'Eating from the wild: Turumbu indigenous knowledge on noncultivated edible plants, Tshopo District, DR Congo', *Ecology of food and nutrition*, Vol. 49, No. 3, pp.173–207.

Thrupp, L.A. (2000) 'Linking agricultural biodiversity and food security: the valuable role of agrobiodiversity for sustainable agriculture', *International Affairs*, Vol. 76, No. 2, pp. 283–297.

Turner, B.L., Kasperson, R.E., Matson, P.A., McCarthy, J.J., Corell, R.W., Christensen, L., Eckley, N., Kasperson, J.X., Luers, A., Martello, M.L., Polsky, C., Pulsipher, A., & Schiller, A. (2003) A framework for vulnerability analysis in sustainability science. *Proceedings of the National Academy of Sciences*. 100(14), pp. 8074–8079.

UNEP (2012) *Measuring Progress: Environmental Goals & Gaps*, United Nations Environment Programme, Nairobi, Kenya.

United Nations [UN] (1987) *Report of the world commission on environment and development: 'Our common future'*. New York, United Nations.

United Nations [UN] (1996) *Rome declaration on world food security*. Rome, FAO.

UN-Water. (2014) *The United Nations Water Development Report 2014: Water and Energy*, UNESCO, Paris, France.

WHO (2008) *WHO European Action Plan for Food and Nutrition Policy 2007–2012*. World Health Organization Regional Office for Europe, Copenhagen.

Part III

Impacts and outcomes of sustainable food system assessment

8 Building the foundation to grow food policy

The development of a toolkit to measure advocacy capacity

Anne Palmer and Raychel Santo

Introduction

In order for social change to be effective and long lasting, change needs to occur at the political, social, and economic levels (Economos et al., 2001). Programmes and services directed at changing human behaviour can influence social norms, but those norms need to be reinforced by a supportive policy environment to achieve scale and sustainability (Stachowiak, 2013). Advocating for policy change helps to create a supportive policy environment for programmes to operate successfully (Huang et al., 2015). Ideally, programmes and policy mutually support one another, but in practice, this convergence is challenging (see also Chapters 2, 5, 6, & 11, this volume). Advocacy can help to overcome that challenge and is a critical tool for food policy councils (FPCs) to utilize.

Broadly defined, advocacy is any activity that aims to shape political, social, and economic outcomes in government and society (Reid, 2000, p. 6). Organizations use various methods to advocate for issues such as mobilizing and training community members, conducting public education, using mass and social media to change social norms, pressuring companies and corporations to enact socially responsible policies, registering voters, and conducting research. Lobbying is a form of advocacy that is directed at influencing policymakers or the public to support or oppose a specific piece of legislation (Harmon et al., 2011). Many FPCs are reluctant to lobby because of its inherent political nature, however, lobbying is an important tool for any group advocating for social change (Chen et al., 2019).

This chapter is specifically focused on how FPCs can build their capacity to advocate effectively for their issues (see also Chapters 2, 4, 5, 9, 10, & 11, this volume, for more on capacity building). They have opportunities to shape public policy, particularly at the local and state levels where relationships with policymakers may have a more immediate influence. FPCs are groups that engage diverse stakeholders to address food systems-related issues and needs within a specified jurisdiction, primarily through policy. Our definition of policy is broad and includes laws and ordinances;

how policies get administered, funded or implemented at local, state, tribal/First Nations, or federal levels of government; as well as changes in institutional (e.g., schools, hospitals, government agencies) practices. Policy work could include working directly to change these various policies, as well as educating or coordinating others who might be advocating for such policies. Other groups that have chosen a different name, such as food council, network, alliance, coalition, committee, collaborative, or partnership, may also fit this definition and benefit from the toolkit. We choose to refer to them as FPCs because that is the most commonly used term[1] to describe such a group in North America.

FPCs organize at the local (city or county), regional, state/provincial, or Native American/First Nations levels[2] to discuss, shape, and assess food system policies and programmes in their communities. While many exist as grass-roots coalitions independent of government, they may also be sanctioned by a local or state government body. Heterogeneous in structure, membership, and issue priorities, they share a collective desire to reform food system programmes and policies through strategic partnerships. While many FPCs aspire to influence policymakers, they may lack knowledge about how government works, whom to approach, and how to frame their issues, all of which hampers them from achieving their advocacy and policy goals. Others strategically decide to avoid policy work because of political realities in their communities.

Recognizing these common challenges, we created a comprehensive online self-assessment toolkit that helps stakeholder groups like FPCs reflect upon their capacities to influence local and state level[3] food policy in order to identify how and where they can build upon them. The toolkit elicits responses to a number of indicators that reflect the specific activities or capacities that each FPC may have. It also provides a sequence of activities to help FPCs better understand the advocacy and policy process, evaluate their current advocacy capacity, or use the results to guide discussions about how to get started. While the toolkit was designed to assist FPCs embedded within the North American policy context, FPCs and similar entities in other industrialized countries (CLF, 2018b) may also benefit from employing it with their groups. With the exception of indicators on specific policy actions that groups may take – which local, regional or state governments in other countries may have different authorities over – the indicators on organizational leadership, decision-making, strategy, and communication are all integral to efficacious organizational operations even beyond advocacy.

This chapter discusses the development of the toolkit; its goals, objectives, and contents; and an example of how it has been used. In the following section, we describe how the toolkit was grounded in Kingdon's (1995) policy windows theory, an approach that encourages advocates to both create and take advantage of opportunities to promote their policy issues when certain conditions are met. In the subsequent sections, we describe the goals, development, and content of the toolkit, with specific attention to equity and systems-thinking metrics. We then reflect on the experience of one FPC that used the toolkit to assess their readiness to advocate. We conclude by discussing how the process of creating this toolkit could inform other efforts aiming to measure the impact of local food initiatives on local food policies.

Theoretical foundation

Several theories of change exist that explore how policy change happens and the effectiveness of specific advocacy tactics, from the 'Large Leaps' (Baumgartner & Jones, 1993) and 'Power Elites' (Domhoff, 1990; Mills, 2000) theories at the global level to tactical theories around 'Messaging and Frameworks' (Tversky & Kahneman, 1981) or 'Diffusion of Innovations' (Rogers, 2003). Stachowiak (2013) presents a summary and comparison of ten different theories of change. By assuming a proposed policy strategy, one is selecting a theoretical approach that will undergird the design and implementation of advocacy efforts. We selected John Kingdon's (1993, 1995) policy windows theory to inform the development of the toolkit because it accounts for the opportunistic nature of policy change, while not discounting the need for a strategy.

This theory suggests that certain conditions foster a 'policy window' to attract policymakers' attention: (1) The issue needs to be identified by policymakers as a serious issue that warrants intervention; (2) Ideas need to be generated about potential solutions to the problem that are feasible, supported by the public, affordable, and reflect commonly held values among policymakers' constituencies; and (3) Political factors such as who is in office, the current political climate, and the influence of the opposition are also considered. According to this theory, at least two of these conditions need to be in place in order to create a 'policy window' – that is, an appropriate and effective time to introduce a new policy.

First, a condition needs to be elevated to a level of concern to be considered actionable. Conditions may garner public attention because of publicized research findings, advocacy campaigns by alliances, or natural or human-caused disasters. Some conditions transition to problems when the public perceives the issue as contrary to public values and subsequent attention shifts towards solving the problem. Next, policymakers propose how they would approach policy discussions to address the problem. When the problem moves into the proposal phase, policymakers inquire about the feasibility and associated costs of the potential solution as well as the public's perceptions of potential solutions. Policy proposals may be influenced by research, but researchers are not the only experts whose advice needs to be heeded. Finally, politics of the problem will be factored into whether or not a policy gets support. When the problem, proposal, and politics align, there is the strongest chance for policy change. According to Kingdon (1993), being positioned to respond to a policy window is more important for advancing policy change than gathering substantial empirical evidence that may or may not influence policymakers.

Many FPCs follow a policy windows approach to their work, although they likely do not identify it as such. The genesis of FPCs' work clusters around food system themes including food access/security, economic development, environmental sustainability, health, food justice/equity, community foods infrastructure such as food production, food processing, food distribution, food procurement, and food recovery/waste. Members may have different perceptions of the problem(s) informed by their world views. Ideally, they gather to understand how solutions need to consider their various perspectives and positions, although it should be acknowledged that some are involved to protect their self-interests. In these early stages, their work may focus on building social and commercial networks, collecting and reviewing existing data, sharing knowledge of the issues, and conducting assessments of their current local or regional food system. This enhanced understanding can help all members appreciate the dimensions of the various issues and how to position those issues to attract support. Significant time, resources, and energy are needed to move a problem into a policy opportunity.

Most FPCs organize themselves into working groups divided by food system sector (e.g., food production, food waste/recovery), function (e.g., fundraising, communications), or both. Working groups assess problems and actions that are likely to have the most influence in their sector. Policy may be one of those actions. Regardless of what policy or policies they focus on, they undergo a process to determine the feasibility of success. Ideally, an FPC will have several options it is ready to propose, when and if the political conditions are deemed appropriate; hence, they will seek to open policy windows as opportunities arise.

One hallmark of most FPCs is diverse membership, with representation from actors across the supply chain and those influencing the sectors, such as government, civil society, and academia representatives. Many FPCs, even those that are not embedded within a government agency, include government staff (79 per cent of US FPCs) or elected officials (30 per cent of US FPCs) as part of their membership (CLF, 2018a). These partnerships with policy experts provide legitimacy and visibility for the FPC, as well as insights as to whether the political context is – or could be – supportive for policy change (Clayton et al., 2015). Those relationships are also vulnerable when political leadership changes. The most effective groups also spend significant time allowing members to develop relationships with each other. These internal relationships can help FPCs hone their policy agendas by providing a diversity of member opinions about the logistical and political feasibility and impacts of potential solutions (Clayton et al., 2015) and may help groups weather leadership changes. Members educate one another about respective challenges and reach consensus on which policies to move forward (see also Chapter 7, this volume). In practice, this process can take years and leans heavily on visionary leadership, in-kind contributions, strong relationships with existing organizations and communities, and a strategy or plan as to how the group will move forward. Many FPCs also include (though not always successfully) citizens who are most impacted by food system challenges and potential solutions (McCullagh & Santo, 2014). Their on-the-ground experiences can provide meaningful 'reality checks' for FPCs as they are designing and implementing policy changes.

Goal of the toolkit

The Johns Hopkins Center for a Livable Future's Food Policy Networks (FPN) project began in 2012 to build the capacity of local, state, regional, and tribal groups that seek to influence food policy in their jurisdictions. By hosting a listserv of nearly 1,500 members, the FPN project maintains a virtual network of individuals and organizations that shares resources, success stories, and challenges in order to support the greater community of practitioners. The FPN website also collates a database of 1,200 resources, compiles a directory of all the food policy groups in the US and Canada, and engages researchers via a research forum. Project staff provide in-person and virtual technical assistance to groups around the country. These engagement methods allow staff to gauge the struggles FPCs experience while working on policy and tailor FPN's work to fit their needs. FPN's annual census survey collects information on FPCs' demographics, structure, funding, achievements, and challenges. Through this census, FPN has learned of numerous FPC policy accomplishments, ranging from mobile meat processing ordinances and farm-to-institution policies to acquiring funds for farmer trainings and anti-hunger programmes (CLF, 2018c). Notably, the longer a council has existed, the more they are likely to engage in a variety of policy activities, from submitting testimonies and providing policy recommendations to supporting or directing an advocacy campaign (CLF, 2018a). Groups also frequently mention that they are unsure of how to engage in advocacy work. In fact, policy training/guidance was listed among the top three technical assistance needs reported by FPCs from the 2013, 2014, and 2015 surveys (CLF, 2016). It was because of these findings that the Food Policy Networks project decided to create the 'Get It Toolgether: Assessing Your Food Council's Ability to Do Policy' toolkit (Palmer & Calancie, 2017).

The toolkit has a lot to offer FPCs and their members. Given the variety of sectors and experiences that FPC members represent, they may have little previous exposure to policy work. Thus, educating members on the advocacy and policy process is critical to optimizing their participation. Education can take many forms including informal conversations with other council members, presentations by local or state government staff on how policy changes are made, guided discussions on why advocacy and policy are important, and opportunities for people to see first-hand what issues stakeholders are facing.

The short-term goal of the toolkit is to provide a foundation for FPCs to better engage in advocacy. This includes identifying who needs to be involved, suggesting steps to develop a strategy, demonstrating how to conduct and sustain advocacy efforts, and increasing FPCs' understanding of how policy is implemented. Using Qualtrics as its software platform, the toolkit's length and scope can be adapted to meet the needs of each FPC. Each section takes less than 15 minutes to complete and participants are provided a cumulative score at the end of each section based on their responses. Once the survey has been submitted, participants are emailed their scores with a corresponding set of resources to increase their capacity in specific areas of need. The scores provide a metric that FPCs can use to identify their strengths and weaknesses. By completing the assessment, groups improve their understanding of the policy process and their group's assets and challenges, thereby increasing their capacity to advocate for policy change. Over the long term, groups can use the toolkit to monitor their performance and assess progress in areas in which they have concentrated resources over several years. The FPN project will use the responses to track FPCs' evolution and capacity over time and as a data source for programming decisions.

Process of developing the toolkit

FPN staff began developing the toolkit by conducting an extensive search for other policy evaluation resources and tools that could be adapted for this purpose. The Alliance for Justice's (AFJ) Bolder Advocacy project, which provides expertise and information to non-profits and foundations to support their engagement in advocacy, was particularly impressive. They created the Advocacy Capacity Tool (ACT) as a guide for non-profits engaged in advocacy (Alliance for Justice, 2018). Also available online, the ACT guide provides real time scores that give a numerical measure of groups' advocacy capacity, and specifically, their readiness to engage in advocacy work.

In addition to using ACT, FPN partnered with Larissa Calancie, a Postdoctoral Fellow at the Center for Health Equity Research at the University of North Carolina Chapel Hill, who developed a self-assessment for FPCs as part of her dissertation (Calancie et al., 2017). We adapted her survey for the toolkit's first section on organizational assessment. We also reviewed tools and surveys specifically designed to measure equity and inclusion and adapted questions for our purpose (University of Virginia, 2010; Curren et al., 2016; Public Health Law Center, 2018). Finally, we had representatives from nine organizations – Greater Cincinnati Food Policy Council, Jefferson County (Colorado) Food Policy Council, Lehigh Valley Food Policy Council, Memphis Tilth, Omaha Food Policy Council, Prince George's County Food Equity Council, Public Health Law Center, United Way of New York City, Virginia Food System Council – pre-test the toolkit and make recommendations for clarification and relevancy. Most of their suggestions were included in the final version.

Contents of the toolkit

The toolkit is divided into six sections: (1) organizational assessment; (2) advocacy goals, plans, and strategies; (3) conducting advocacy; (4) advocacy avenues; (5) organizational operations to sustain advocacy; and (6) policy implementation (see also Chapter 9, this volume, about the development of a City Region Food Systems toolkit). Below, we provide a brief description of each section of the toolkit, along with example indicators. Because some FPCs may also support other organizations' advocacy efforts, sections two through six offer a 'relies on partners' response option, which is not scored but noted. 'Relies on partners' means that the FPC has determined that they do not need to embark on this activity themselves and have identified partner organizations on which they can rely, or in some cases, support.

The toolkit can be completed by the organizational leader or facilitator, working group leaders, or by many members of the group or network. For example, it was rolled out among a network of FPCs in Michigan in fall 2018. The respondent's role(s) in the group is identified as part of the demographic information collected with each survey. As the name implies, the toolkit is a tool to gather different perspectives on the readiness of a group to engage in advocacy work; the diversity of those perspectives creates a profile of what members consider to be the group's strengths and weaknesses and can be used to guide a group discussion about next steps.

The first section of the toolkit, on organizational assessment, has three sub-sections: leadership; structure and membership; and networking and relationships. We use indicators such as receptivity to new ideas; decision-making processes; and creating an organizational climate that welcomes participation, provides leadership opportunities, and helps resolve conflict. The structure and membership indicators stress diverse membership and identify steps that enhance the functionality of the group such as by-laws and working groups. Networking indicators seek to determine the connectivity and perceived value of networking among members.

Section two concentrates on advocacy goals, plans, and strategies. It covers three topics related to this theme: preparedness; food policy agendas, plans, and strategies; and adaptability. Preparation acknowledges the need for the group to engage its members in creating a vision that is regularly communicated to all stakeholders. As part of those goals, FPCs should understand the potential policies, priorities, and environment in which they work. When they delve deeper into this process, they may conduct a policy scan or talk to other stakeholders, especially people affected by the issues. The final section includes indicators that consider knowledge of the power structures that influence the policy issue as well as monitoring schemes to assess progress.

The third section on conducting advocacy is the most robust section and explores what research the FPC uses, how they build the capacity of their members to conduct advocacy, whom they partner with to achieve their goals, and how they use communication activities to support their work. Research and analysis questions inquire about their data sources: from whom they get data, if they collect their own data and what methods they use, how they verify accuracy, and collaborative partners.

Section four assesses the FPC's administrative, institutional, and legislative advocacy skills, knowledge, and actions. Administrative advocacy (Alliance for Justice, n.d.) refers to actions related to rules, regulations, and other administrative actions that are not specific to legislation. Institutional advocacy refers to actions within government or a private institution. Legislative advocacy refers to actions that take place in legislative bodies such as a municipal council or state congress rather than other government bodies. The questions assess the extent to which the organization understands the processes involved in advancing these different types of policies; works on – or supports partners working on – the development of such policies; and serves as a resource for policymakers.

Section five reviews the organizational operations that support advocacy such as leadership's understanding of regulations about advocacy and lobbying, investment in training for members to do advocacy, and involvement in passing policies. Ability to fundraise for advocacy activities has been a challenge for many FPCs and this section also includes a number of metrics that assess relationships with funders, ability to obtain support for their work, and financial management practices.

Section six focuses on policy formulation, enactment, outputs, and outcomes. Indicators assess if members have organized community members to advocate, and whether they have provided testimony in support of or against any policies, met with policymakers, and developed monitoring mechanisms. In addition, this section asks about whether or not there has been an increase in awareness of food policy issues among various audiences, and whether or not the group is perceived as a resource for policymakers. A long list of possible food policy outputs is intended to expose participants to the variety of policies that groups can work on. Finally, we ask groups to speculate about how the policies they have worked on may have contributed to desirable outcomes in their communities.

Evaluating equity and inclusion

In addition to thematically organizing the toolkit into sections based on the steps of engaging in policy advocacy work, we integrated into all sections of the toolkit two themes that we believe should be central to the work of all FPCs: (1) equity and inclusion and (2) systems thinking. Table 8.1 highlights the metrics we selected for groups to determine if they are implementing their work with a lens on equity and inclusion. Equity was explored in terms

Table 8.1 Indicators assessed in each section of the toolkit

Section	*FPC features assessed*	*Equity and inclusion metrics*
Organizational assessment	• Leadership, group structure and functions, membership recruitment, engagement, networking	• The organization promotes and supports diverse representation and participation on the council; provides opportunities for members to build leadership within the organization; and shares power in decision-making with the organization's members. • The organization adequately reflects the racial, economic, gender, and ethnic diversity of the community it represents.
Advocacy goals, plans, and strategies	• Clear, relevant agenda that defines food policy goals, prioritizes activities, and reflects community needs • Flexible plan to carry out policy priorities	• The organization regularly provides opportunities to hear about food-related issues from community members who are not on the council. • The organization partners with community groups to increase and promote community engagement in local decision-making, particularly in low-income and historically marginalized neighbourhoods. • The organization has considered how the proposed policy agenda will impact socially disadvantaged and marginalized groups.
Conducting advocacy	• Organization researches policy issues and good practices • Capacity building of members to work on policy issues • Partners with other organizations and decision makers to advance policy goals • Communication strategy and media engagement	• The organization uses surveys, focus groups, or other research methods to better understand community interests, needs, or concerns about a specific policy issue. • The organization identifies segments of the public to educate about its agenda. • The organization implements a plan as needed to expand the size and diversity of its membership to achieve policy objectives.
Advocacy avenues	• Organization's skills, knowledge, and actions related to administrative, institutional, and legislative advocacy	• None specific to equity and inclusion.

(*continued*)

Table 8.1 (Cont.)

Section	*FPC features assessed*	*Equity and inclusion metrics*
Organizational operations to sustain advocacy	• Organizational commitment, funding advocacy, and decision-making structure indicators	• The organization invests in building the capacity of its members to strengthen its advocacy work. • The organization has increased its budget over time.
Policy implementation	• Policy formulation and enactment	• The organization has mobilized community members to advocate on behalf of a priority policy issue. • The organization engaged community members in forming policy statements. • The organization has included community members to help with policy adoption, implementation, or evaluation processes.
	• Policy outputs	• The organization has worked on policy changes that support living wages. • The organization has worked on policy changes that improved labour conditions. • The organization's actions have provided financing or credit for people who would otherwise not have access (for food-related enterprises). • The organization has worked on or supported organizations that are working on policies that address economic or housing development and food access simultaneously.
	• Policy outcomes (have contributed to an increase in ...)	• The wages of food systems workers • (Improvement in) the state of working and living conditions for food or agriculture workers • Jobs for people that have had employment challenges • Access to credit or capital for people who would not have had access through traditional means • Value-added processing facilities that provide economic opportunities to those who need them.

of how the group considers the effect of policies and programmes on communities of colour, people living in poverty, indigenous groups, rural communities, (im)migrants, and youth and how it elevates the power of those groups to participate in food system changes. As mentioned previously, the survey can be completed by group leaders or by many members, which elicits various perspectives in the organization's assessment.

Systems-thinking metrics

In addition to equity measures, we included metrics that would nudge groups towards systems thinking. Systems thinking can be defined as 'an enterprise aimed at seeing how things are connected to each other within some notion of a whole entity' (Peters, 2014). Complex issues such as those that affect the food system, are best understood when using systems thinking and systems approaches (Clancy, 2014b, see also Chapter 4, this volume). Systems thinking acknowledges that solutions will require engagement from more than one sector/organization; considers long-term, short-term and unintended consequences; identifies leverage points that could lead to change; and considers how things change over time and accounts for tracking changes. Many of these concepts are allied with the principles of FPCs, although members may articulate them differently. We hope that by making these concepts more explicit, groups will be more comfortable in the application of systems thinking throughout their work.

Even if and when FPCs apply systems thinking, it is worth acknowledging that they may encounter difficulties when attempting to tackle multidimensional food systems issues within conventional policy paradigms. Most food issues are typically addressed in silos within traditional policy sectors (e.g., production, economic development, health), which makes advocacy on systems issues more diffuse and complex.

Toolkit in action: a reflection on one FPC's experience

Since the toolkit was released in winter 2017/2018, at least 70 people from 30 FPCs have completed it. These numbers were too small to analyse the toolkit's impact on practitioners at a quantitative level. Thus, we decided to speak with three representatives from one of the toolkit's early adopters, the Lynchburg Area Food Council (Virginia), to provide some initial qualitative perspectives.

Background

The Lynchburg Area Food Council (LAFC) formed in 2012 with the goal of collaborating on tangible programmes that would support the community across several counties in central Virginia. Informant 1, a charter member and vice president of the LAFC, was involved in the council's early efforts, which

Table 8.2 Systems-thinking considerations

Section	*Systems-thinking considerations*
Organizational assessment	• The leadership is receptive to new ideas. • The organization promotes and supports diverse representation and participation on the council. • The organization adequately reflects all food system sectors (producers, policymakers, food businesses, public health, etc.). • Joining the organization has helped coordinate efforts among various organizations that other members belong to or represent.
Advocacy goals, plans, and strategies	• The organization understands the overall policy environment related to its priorities. • The organization gathers information and recommendations from constituents and other stakeholders in the development of its food policy agenda. • The organization has considered how the proposed policy agenda will impact socially disadvantaged and marginalized groups. • The organization has some monitoring mechanisms in place to help assess progress and make course corrections when necessary.
Conducting advocacy	• The organization seeks guidance from other organizations and stakeholders to understand their policy priorities. • The organization identifies stakeholders (outside of its membership) that have similar goals, including those with complementary knowledge/skills, with which it could collaborate on policy. • Outside of its membership, the organization seeks support from stakeholders who may not be traditional allies, but with whom it could partner on a specific policy issue.
Advocacy avenues	• N/A
Organizational operations to sustain advocacy	• N/A
Policy implementation	• The organization has monitoring mechanisms in place to know whether or how the policy is being implemented. • The organization's actions have led to an increase in awareness of food system issues among the FPC members, elected officials, or general public.

concentrated on food access and food security research to identify food deserts. In response to those efforts, Randolph College conducted a community food assessment with residents living in low-income neighbourhoods, and United Way played a facilitative role for the council. Over time, the LAFC has shifted to focus more on local food, small farmers, sustainable production, and agricultural extension activities and less on food access, primarily due to membership changes. For example, after the local health department received a grant to construct a garden, LAFC also began offering small garden grants to communities.

At one of their bi-monthly meetings, Informant 2, one of the council members, introduced the toolkit to the group as a way for them to gather information to decide what they should focus on next. The timing seemed fortuitous. Informant 1 and Informant 3, an employee of the Virginia Department of Health, noted the group's readiness and desire to move activities forward. Every member was encouraged to complete it (8 out of 10 of them did), so that the results would include a range of perspectives.

What they learned

Once LAFC completed the toolkit, Informant 2 discussed her realization that 'we have been limited only by the fact that we didn't know all that was possible. That's one benefit of completing the toolkit, it gave us ideas we may have not thought of before.' While they excluded the term 'policy' in their council's name, she recognized that they could still work on policy; they just needed to consider how they could manoeuvre into the policy arena. As a group of community members, the LAFC aspired to work with city council members. Answering the toolkit's questions helped them acknowledge that in order to do that, they needed to identify specific policy partners. The section that helped them review their tangible accomplishments would further build momentum to work on policy.

The toolkit also helped the LAFC members realize that they had not taken time to think about their current role in the community and what they would like that role to be. Completing the toolkit encouraged members to think about those issues and use the results as a discussion starter. The scan of potential food and agriculture policies that FPCs could work on also prompted them to think about the council's potential to serve as a resource to gather information to share with the public, and more specifically, how they might be more of an asset to the community. As Informant 3 shared,

> we need to be creating knowledge and educating. We haven't been amassing a group of people who care about this work and what we do. Having a communications strategy would help us to communicate with the public. I don't think many people know we exist right now.

Informant 1 valued how the process of completing the toolkit gave members time to reflect on broader questions such as what the LAFC's overall goal is, what their individual roles as council members are, and how they make things happen. 'The toolkit asks good questions as to what your leadership looks like, what difference are you making in social change or policy? Have we done enough in our local city council to have a voice?' He also remarked that while they have made several attempts at short- and long-term planning, the toolkit experience highlighted how asking the right questions and listening to the answers may be more important.

The results illuminated points of convergence and divergence among members, such as whether or not the LAFC's representation reflects the diversity of the community they support. Some people thought it did, others not so much. As Informant 2 contemplated, 'Is it important for us to have agreement about these issues? How do we come together and answer those questions?' Where there were areas of discrepancy in the responses, they realized they would like to have a better idea of how to approach such differences in opinion. In its list of additional resources, the toolkit does include two documents that provide guidance on problem-solving and decision-making in groups (CLF, 2017).

The toolkit also revealed opportunities for enhancing the LAFC's organizational structure. Informant 3 described their group as relatively informal and embraced several suggestions about policies and procedures that they could work on to improve the group's operations. For instance, she noted that they have no orientation for new members and thought they could do more activities to build relationships among members. Such activities could help members learn from one another, 'We need to expand our horizons and group learning; members need to be educated about other [food] issues.'

As with many FPCs, the Lynchburg Area Food Council tries to balance its time and resources to meet community needs. Informant 1 described it as the tension between 'Are we about programs, or are we about systems change and advocacy and policy – the 30,000-foot view?' As Informant 3 reflected, the toolkit helped them assess where they are right now and where they might go in the future. 'We've got a lot of work to do, and a big future ahead and now we have a target line as to what we can be shooting for.'

Conclusion

We embarked on developing this toolkit as a way to provide FPCs with an easy means of assessing their current capacity to advocate on behalf of their food system issues. Time and again we found themes from the FPN trainings, listserv conversations, conferences, surveys, and research, which indicated that people were looking for a how-to guide to begin their advocacy work. By necessity, many groups spend a couple of years in development before they actually begin identifying a 'policy' issue that begs for attention. Once they arrive at that place, their next steps are not always clear. The toolkit presents an opportunity for any group working on food system change to better understand the advocacy process and what needs to be in place to optimize their resources. This learning occurs at various scales. The people using the toolkit get feedback on their status and can periodically repeat the assessment to measure change over time. Groups can also use the results to make decisions about how their resources are allocated, who else needs to be engaged in the work, and what strategies they could employ to improve their outcomes. Simultaneously, FPN project staff use the data to determine what additional materials, webinars, or other activities can be offered to support groups. Moreover, aggregated results from the toolkit can be shared with other food policy groups working on food system change to transfer knowledge among groups at the local and state levels (Clancy, 2013). All of this feedback helps to avoid repetition of mistakes.

Another concept that is covered in the toolkit is governance: 'managing, steering and guiding of public affairs by governing procedures and institutions in a democratic manner' (Pisano et al., 2011, p. 3, as cited in Clancy, 2014a). Many FPCs structure themselves to influence public or institutional policy, sometimes both. By building relationships with actors in the various food sectors, identifying government departments that influence particular administrative actions, examining legislative options, and inviting institutions into that process, they encounter short- and long-term policy opportunities. As Informant 2 from Lynchburg noted, the toolkit 'made me realize that we were limiting ourselves to what we thought we could do'.

When many FPC members from the same organization complete the assessment, the points of convergence and divergence among member perspectives become obvious. This heterogeneity reflects a feature of systems thinking (Clancy, 2014b). More important than any score, if facilitated effectively, these differences provide insights into various world views and a starting place for rich discussions.

When the group agrees to allocate time to using the toolkit, they may also want to consider how they will manage survey administration and interpretation of results. FPN has held phone consultations with a few groups as well as compiled the scores and shared the raw data, in the event they want to do any analysis. It may be appropriate to invite an outside facilitator to help the group process the results, particularly if dramatic differences emerge. Taking time to discuss those differences will test the trust among group members. Since most groups use consensus-based decision-making, these discussions can help members understand conflicting views and discuss promising places for policy change. How to form a specific policy position among a diversity of member perspectives is still a challenge that many FPC leaders face, but maintaining a place to have those conversations is the first step to overcoming such divergences (Santo & Moragues-Faus, 2018; see also Chapter 6, this volume).

The final step, which is never really final, is deciding what's next. It would be easy to get overwhelmed by the volume of questions and subsequent scores. Given the paucity of resources for most groups, prudence in moving forward is highly recommended. FPN created the list of resources for precisely that reason – to give people a limited amount of information that they can choose to use if they want to improve their ability to work on a particular issue. When possible, each question was considered a discrete metric and a companion resource was identified for that metric. Most of these resources already existed in the FPN resource database. Some groups may not search the resources to answer their questions, instead using the process of doing the toolkit as a starting place to collectively answer the question, 'what's next?' Deciding to explore further, with resources in hand, can improve the efficacy of any group's work.

The toolkit can help FPCs pause and reflect, a process that is difficult for individuals, let alone when one is part of a group. A group may decide to use it annually, as a way to measure any change. Or they may only focus on one section with the goal of improving their performance in a particular area. The goal is to provide a starting point for all the members to think about broader goals and objectives. Reflecting on their work forces them to think and rethink about how they are adding value to the greater good of creating a healthy, sustainable, and fair food system. Many groups hesitate to explicitly state that they work on policy (Schiff, 2008, p. 211; Santo & Moragues-Faus, 2018, p. 10), but they might be underestimating their ability to work on policy issues. Policy is just a way to change standard operating procedures, whether they exist in legislation, within government administrative actions, or in private institutions. FPCs are more frequently mentioned as a means of mobilizing at various levels to affect food system change, however, that recognition has not attracted commensurate resources to match the interest. With diverse partnerships, adequate resources, and training, their capacity to actively engage in advocacy and policy can increase substantially. FPN aspires to use the results from the toolkit to improve the content, expand the indicators on systems to a more sophisticated level and, if groups choose to use it on a regular basis, compare their change over time. These metrics shape our understanding of how policy can be driven by food policy groups.

Acknowledgements

Thanks to the members of the Lynchburg (VA) Area Food Council, Liza Dobson, Lindsay Cawood, and Pat Haley for being interviewed for the case study. We would also like to acknowledge Dr. Larissa Calancie, a Postdoctoral Fellow at the Center for Health Equity Research at the University of North Carolina Chapel Hill, who co-developed the toolkit with Anne Palmer.

Notes

1 Of those that have responded to an annual census of food policy groups in the US and Canada, 28 per cent of the 'active' groups in 2018 had names that included the term 'food policy council', 21 per cent used 'food council', 9 per cent used 'network', 8 per cent used 'alliance', 8 per cent used 'coalition', 5 per cent used 'committee', and 21 per cent used other terms (CLF, 2018a).

2 Seventy-one per cent of FPCs in the US and Canada operate at the local level, 20 per cent operate at regional (e.g., multi-county, multi-state), 8 per cent at state/provincial, and 1 per cent at Native American/First Nations levels (CLF, 2018a).

3 It is worth noting that many local and state-level policies are often constrained by national (e.g., Farm Bill) or international (e.g., trade agreements) policies over which local FPCs do not often have significant capacity to influence individually (Clancy, 2012, 2014a). Some FPCs do engage in public education about such policies, and a few are very active in the national policy scene. In theory, local groups could also collectively advocate around higher-level policies, though such activity has been limited thus far (Santo & Moragues-Faus, 2018).

References

Alliance for Justice (2018) *'Advocacy capacity tool', Bolder Advocacy*. Available at: https://afj.co1.qualtrics.com/jfe/form/SV_3g5BupEfjKpkFqR.

Alliance for Justice (n.d.) *'Administrative advocacy: Influencing rules, regulations, and executive orders', Bolder Advocacy*. Available at: http://www.foodpolicynetworks.org/food-policy-resources/?resource=964.

Baumgartner, F.R. & Jones, B. (1993) *Agendas and instability in American politics*. Chicago, IL, University of Chicago Press.

Calancie, L., Allen, N.E., Weiner, B.J., Ng, S.W., Ward, D.S., & Ammerman, A. (2017) Food policy council self-assessment tool: Development, testing, and results. *Preventing Chronic Disease*. 14(160281), pp. 1–14.

Chen, C., Santo, R., Ellison, B., & Maiolatesi, A. (2019) *Advocacy and Lobbying 101 for Food Policy Councils*. Baltimore, MD, Johns Hopkins Center for a Livable Future and Harvard Food Law and Policy Clinic. Available from: https://www.chlpi.org/wp-content/uploads/2013/12/Advocacy-for-FPC-toolkit.pdf.

Clancy, K. (2012) Issues of scale. *Journal of Agriculture, Food Systems, and Community Development*. 3(1), pp. 21–23.

Clancy, K. (2013) Feedback loops. *Journal of Agriculture, Food Systems, and Community Development*. 3(3), pp. 5–7.

Clancy, K. (2014a) Food system governance. *Journal of Agriculture, Food Systems, and Community Development*. 4(2), pp. 3–6.

Clancy, K. (2014b) A different way to approach policy change. *Journal of Agriculture, Food Systems, and Community Development*. 4(4), pp. 9–11.

Clayton, M.L., Frattaroli, S., Palmer, A., & Pollack, K.M. (2015) The role of partnerships in US food policy council policy activities. *PloS One*. 10(4).

Curren, R., Nelson, J., Marsh, D.S., Noor, S., & Liu, N. (2016) *Racial equity action plans, a how-to manual*. Berkeley, CA, University of California, Haas Institute for a Fair and Inclusive Society.

Domhoff, G.W. (1990) *The power elite and the state: How policy is made in America*. New York, Aldine De Gruyter.

Economos, C.D., Brownson, R.C., DeAngelis, M.A., Foerster, S.B., Foreman, C.T., Gregson, J., Kumanyika, S.K., & Pate, R.R. (2001) What lessons have been learned from other attempts to guide social change? *Nutrition Reviews*. *59*, pp. S40–S56.

Harmon, G.M., Ladd, J.A., & Evans, E.A. (2011) *Being a player: A guide to the IRS lobbying regulations for advocacy charities*. Washington, DC, Alliance for Justice.

Huang, T.K., Cawley, J.H. Ashe, M., Costa, S.A., Freiches, L.M., Zwicker, L., Rivera, J.A., Levy, D., Hammond, R.A., Lambert, E.V., & Kumanyika, S.K. (2015) Mobilisation of public support for policy actions to prevent obesity. *The Lancet, Obesity*. 3(385), pp. 2422–2431.

Johns Hopkins Center for a Livable Future [CLF] (2016) *Food policy council trends 2013–2015* [Data file and personal communication].

Johns Hopkins Center for a Livable Future [CLF] (2017) *Recommended resources for 'Get it Toolgether: Assessing Your Council's Ability to Do Policy Work'*. Available at: http://www.foodpolicynetworks.org/fpc-map/FPC-around-the-world.html.

Johns Hopkins Center for a Livable Future [CLF] (2018a) *2018 Food policy council census results* [Data file and personal communication].

Johns Hopkins Center for a Livable Future [CLF] (2018b) *Food policy groups around the world*. Available at: http://www.foodpolicynetworks.org/fpc-map/FPC-around-the-world.html.

Johns Hopkins Center for a Livable Future [CLF] (2018c) *Food policy council achievements wheel*. Available at: http://www.foodpolicynetworks.org/food-policy-resources/?resource=1236.

Kingdon, J. (1993) How do issues get on the public policy agenda? In: Wilson, W.J. (Ed.) *Sociology and the public policy agenda*. American Sociological Association Presidential series. Thousand Oaks, CA, Sage Publications.

Kingdon, J.W. (1995) *Agendas, alternatives, and public policies* (2nd ed.). New York, Harper Collins College.

McCullagh, M. & Santo, R. (Eds.) (2014) *Food policy for all: Inclusion of diverse community on food policy councils*. Baltimore, MD, Center for a Livable Future.

Mills, C.W. (2000) *The power elite* (revised edition). New York, Oxford University.

Palmer, A. & Calancie, L. (2017) *Get it toolgether: Assessing your food council's ability to do policy*. Baltimore, MD, Johns Hopkins Center for a Livable Future.

Peters, D. (2014) The application of systems thinking in health: Why use systems thinking? *Health Research Policy and Systems*. 12(51), pp. 1–6.

Public Health Law Center (2018) *Focusing on equity and inclusion when we work on Public Health laws*. Saint Paul, Mitchell Hamline School of Law. Available at: https://www.publichealthlawcenter.org/resources/focusing-equity-and-inclusion-when-we-work-public-health-laws-2018.

Reid, E.J. (2000) Understanding the word 'advocacy': Context and use. In: Reid, E.J. (Ed.) *Structuring the inquiry into advocacy, Volume 1*. Washington, DC, The Urban Institute, pp. 1–8.

Rogers, E.M. (2003) *Diffusion of innovations* (5th ed.) New York, Free Press.

Santo, R., & Moragues-Faus, A. (2019). *Towards a trans-local food governance: Exploring the transformative capacity of food policy assemblages in the US and UK*. Geoforum, 98, 75–87.

Schiff, R. (2008) The role of food policy councils in developing sustainable food systems. *Journal of Hunger & Environmental Nutrition*. 3(223), pp. 206–228.

Stachowiak, S. (2013) *Pathways for change: 10 theories to inform advocacy and policy change efforts*. Center for Evaluation Innovation and ORSIMPACT.

Tversky, A. & Kahneman, D. (1981) The framing of decisions and the psychology of choice. *Science*. 211(4481), pp. 453–458.

University of Virginia, Institute for Environmental Negotiation (2010) *Food Policy Audit*. Charlottesville, VA, University of Virginia. Available at: http://www.foodpolicynetworks.org/food-policy-resources/?resource=271.

9 Tools for food system change

City Region Food System assessment, planning, and policy

Guido Santini, Marielle Dubbeling, and Alison Blay-Palmer

Introduction

While linked to other concepts such as foodsheds, bioregions, short food-supply chains, and territorial food systems, in recent years, the City Region Food System (CRFS) approach has emerged as a way to connect typically divided urban and rural spaces (see also Chapters 2 & 4, this volume). In addition to spatial integration, the CRFS approach also emphasizes coherence across food-chain dimensions, taking into account environmental and socio-economic aspects as part of fostering more resilient and sustainable food systems (for a more detailed analysis please refer to Blay-Palmer et al., 2018 or Blay-Palmer & Renting, 2015). To realize these goals, city regions can apply a large number of strategies and tools, such as the promotion of (peri)urban agriculture; preservation of agricultural land areas and watersheds through land-use planning and zoning; development of food distribution and social protection programmes; support for short supply chains and local procurement of food; and promotion of food waste prevention, reduction, and management. Developing a resilient CRFS, however, requires political will – integrating available policy and planning instruments (e.g. infrastructure, investment, logistics, public procurement, land-use planning); involvement of various government departments and jurisdictions (local and provincial); and inclusive organizational structures at multiple scales (municipal and district among others). Improved CRFSs offer the opportunity to help achieve better economic, social, and environmental conditions in both urban and surrounding rural areas by activating new or reinforcing existing concrete policy and investment opportunities.

In 2015, FAO and RUAF Foundation, in collaboration with the Laurier Centre for Sustainable Food Systems (LCSFS), and with the financial support of the German Federal Ministry of Food and Agriculture, the Daniel and Nina Carasso Foundation, and the CGIAR Water Land and Ecosystems Programme led by IWMI, embarked on a collaborative programme with regional partners to operationalize the CRFS approach. The goal was to assess and plan increasingly sustainable city region food systems in seven city regions around the world with an emphasis on the Global South.

The pilot cities are: Colombo (Sri Lanka), Lusaka and Kitwe (Zambia), Medellin (Colombia), Quito (Ecuador), Toronto (Canada), and Utrecht (the Netherlands) (FAO & RUAF, 2015; Dubbeling et al., 2017).

The CRFS approach builds on a formalized process of identifying and engaging all relevant stakeholders from the start of the assessment process through to policy review and formulation (see Chapters 2 & 7, this volume). This means that a CRFS process can result in revised or new urban food policies, strategies, and projects, and also in the creation of new – or revitalized – networks for food governance and policy development, such as urban food policy councils and new institutional food programmes and policies (Blay-Palmer et al., 2018).

Following the CRFS research, the approach was translated into a toolkit that provides guidance on how to assess a CRFS and then helps to build a more sustainable city region food system (the methodology is available at http://www.fao.org/in-action/food-for-cities-programme/toolkit/introduction/en/). The toolkit was developed out of, and is supported by, three phases of research that produced synthetic reports covering each pilot project. These phases involved data identification and consolidation from existing sources, and then the generation of new data followed by policy assessment and recommendations. The toolkit is available online (http://www.fao.org/in-action/food-for-cities-programme/toolkit/en/) as a set of linked documents that guide users through an iterative, non-linear process, including non-sequential phases to establish multi-stakeholder task forces, develop a vision, collect data, identify areas for improved food system sustainability, and work towards policy coherence. The toolkit includes more than 40 tools and resources – such as meeting guidelines and policy examples – developed and identified during the pilot phase. These examples from the cities offer an overview of why and how each city region implemented their changes and what outcomes each achieved. It is meant to be a resource for policymakers, researchers, and other key stakeholders who want to better understand their own CRFS and plan for improvements. In this way the examples and tools provide valuable experiences, expert guidance, and lessons that may accelerate the development of similar initiatives in other city regions that wish to apply, customize, or scale up similar practices.

This chapter provides an overview of the research results from the pilot cities that inform the CRFS toolkit as well as concrete examples that illustrate how pilot city regions adopted and adapted their strengths and peculiarities of their own contexts to generate and share evidence that led to policy outcomes. In particular, the chapter delves into the experiences of the pilot city regions that include Colombo (Sri Lanka), Lusaka and Kitwe (Zambia), and Medellin (Colombia). We also include brief overviews of the work in Toronto and Quito.

The CRFS process in the Colombo region triggered policy discussions beyond the local level that are spreading into provincial (regional) and national levels. It created the basis to start visualizing the importance of a territorial approach to food systems and the actions needed to offset the impacts of natural resource management challenges, climate change, and shocks on city regions. It shed light on food safety, nutrient safety, value chain management, and food waste and losses in relation to urban spaces (FAO, 2018). In the two Zambia regions, the CRFS approach raised awareness and political momentum to reinforce the role of horticulture to promote diversified food production and sustainable consumption through joint planning specifically as proposed for the Urban and Regional Planning Act and could also be formative for the ongoing National Urbanization Policy. In the Medellin case, the CRFS approach enabled key policymakers, planners, and practitioners to move from a singular focus on urban food security and nutrition to a more integrated food systems vision that was applied across the region. This facilitated the identification and formulation of territorial planning strategies that strengthen more sustainable and resilient food chains from production to consumption. In turn, this enabled cooperation and coordination of the actors, including new forms of integration and collaboration between producers, agents, and markets, for example, the construction of a closer and more equitable relationship between rural and urban areas, which was designed to meet the needs of the urban and rural areas, producers, and consumers.

The CRFS toolkit and approach

Building from the experiences in the pilot cities, the toolkit was developed to include different considerations typically needed to support a CRFS assessment and planning approach (A diagram of the City Region Food System Approach is available at: http://www.fao.org/in-action/food-for-cities-programme/toolkit/introduction/en/).

The phases in the CRFS process are iterative and not intended to be linear. Consistent with other place-based research, the entry points should be defined based on the local contexts including available evidence and information, capacity, stakeholder engagement, and existing policy agendas (Sonnino et al., 2016). In some contexts, setting the policy agenda could be the starting point and an assessment may be used to explore and assess the policy priorities identified moving back and forth between data gathering and policy development. A city may enter at any point in the process suitable to its local context.

To begin, the CRFS team needs to: engage a multi-stakeholder task force, including researchers, policymakers, and food system participants; establish goals and objectives; and then determine what data and information exist. Typically, this initial phase produces outputs including terms of reference for the project as well as a workplan and timeline. Once the task team has been established, the approach develops based on the needs of the specific city region and would include some or all of the following elements:

- **Defining the nature and boundaries of the CRFS:** A key activity in the defining stage will be to conduct a participatory mapping exercise with a wide range of stakeholders to define the nature and boundaries of the local city region and the city region food system. In many cases these boundaries are based on either available data and/or political boundaries and other administrative considerations. In some cases, boundaries were set based on food-flow considerations as, for example, in Medellin.
- **Characterizing the CRFS – the CRFS scan:** The toolkit provides guidelines about how to map and describe the local city region food system. This includes questions such as: who feeds the city region, where is the food processed, how is it marketed, what do people eat and what is their food security and nutrition status, how is food waste managed and who are the government and institutional actors involved in the food system?
- **Visioning:** The toolkit outlines how to build a shared common vision for a sustainable and resilient CRFS. The vision underpins the different parts of the entire CRFS assessment and planning process. The aim is to build a vision that transcends the given project and can eventually grow into a more refined, consolidated – and political – set of priorities that is agreed upon by all stakeholders involved as the project progresses. The vision gives direction to the implementation of the CRFS assessment and planning.
- **Analysis of the CRFS:** The toolkit suggests ways to analyse current food system performance with regards to different sustainability dimensions, food system vulnerabilities, assets, threats, and weaknesses. The analysis also allows for the identification of opportunities to strengthen the CRFS.
- **Policy planning:** The toolkit provides recommendations for concrete policy and planning interventions in the CRFS and identification of stakeholder roles, (new) institutional frameworks, proposal writing, programmes, and action plans. This may also include the identification of lobbying opportunities and elaboration of specific advocacy materials. The policy support and planning could involve further policy analysis, policy formulation and revision, policy integration, and planning of further action. Continued engagement of policymakers across multiple scales and other stakeholders can be key to ensuring policy uptake and effective implementation.
- **Governance and multi-stakeholder dialogue process:** From a governance perspective, the toolkit presents a CRFS approach that aims to be highly participatory and promotes local ownership of the process through multi-stakeholder, multi-scale engagement as it seeks to foster inclusive dialogue among all the relevant stakeholders involved in the CRFS. The goal is to support local governments and multi-stakeholder bodies in taking informed decisions on food planning and capacity building, recognizing the added value in the consultation–participative processes,

and knowledge sharing. In some cases, the CRFS process has improved food system governance by consistently applying a multi-stakeholder participatory approach and process throughout the various steps of CRFS assessment and planning. This, in turn, can lead to strengthened existing, and the creation of new, networks and/or food governance structures, the improvement of government and stakeholder capacity in implementing a CRFS process, and the promotion of food policy design and monitoring.

The tools, material and examples provided on the toolkit website need to be adapted to the specific circumstances and interests of a city region through the creation of a local CRFS team that can use the toolkit according to local concerns and capacities. Examples of this are provided later in the chapter through case-study work. A sound CRFS process takes into account existing and specific agronomic, economic, and institutional–political conditions; the variety, interests, and expertise of the different involved stakeholders; available resources, existing data and information; and specific set goals in the local context (see Chapter 8, this volume, for other examples of toolkits leading to change).

Outcomes of the CRFS assessment and planning process in pilot city regions

Considering that each city region has its own context, the toolkit is not meant to represent a guideline that fits all. Instead, it has been designed to be a flexible instrument and to adapt to the characteristics and needs of each context. In that respect, the seven pilot city regions have adapted the approach, building on their specific contexts.

As discussed below, in each of the pilot cities the CRFS process has built more awareness and information exchange about the characteristics and functioning of the CRFS and has created the basis for a common and shared vision of a sustainable CRFS. In each city, the CRFS process has led to a set of key policy proposals and recommendations. In some cities this has resulted in policy or project activity, including new governance structures. In other cities, processes will be carried forward by local stakeholders or under new projects.

Case study 1: Two regions in Zambia – The role of the CRFS approach in raising awareness and political momentum to promote diversified food production and sustainable consumption

Zambia is a landlocked country located in south-central Africa. Forty-one per cent of its population lives in urban areas (urbanization rate in Africa: 38 per cent), mostly gathered in two regions (World Bank, 2016); Lusaka, the capital city (1.7 million inhabitants, Lusaka Statistical Office 2010) and surrounding districts; and the Copperbelt Province, including the city of Kitwe (468,682 inhabitants). With the repetition of droughts and rapid urbanization patterns, both regions face diverse challenges to ensure food security and nutrition for all, while providing decent livelihoods to farmers and making efficient use of natural resources.

In addition, the existing food system in Zambia, built on large-scale mono-cropping of maize, is eroding ecosystems and crop diversity and reducing diversity in consumption and diets. Today, the Zambian diet is mainly composed of cereals, predominantly maize, starchy roots and, to a lesser extent, fruits and vegetables. Cereals provide almost two-thirds of the dietary energy supply. According to the National Food and Nutrition Commission (NFNC), one of the major causes of a high rate of malnutrition in Zambia is the mono-diet practice. The culture of mono-diet is born from mono-cropping food production, which is heavily slanted towards maize (Lusaka Central Statistical Office, 2010; Lusaka Government, 2015; Biriwasha, 2017).

In both city regions, prior to the CRFS pilot project, very little had been done to either analyse or plan the food system. As a result, few data and studies were available, and looking at food through a system and multi-stakeholder lens was still at a very preliminary stage. Despite the existence of a Ministry for Local Governments, food supply and distribution is still mainly handled by the Ministry of Agriculture. Both city councils showed interest in taking up this approach providing the needed political buy-in to initiate the process.

In both city regions, a multi-stakeholder group was formed including: producers, supermarkets, marketeers, processors, cooperatives, ministries, NGOs, municipalities, and consumers associations. The city regions were defined based on administrative boundaries and food flows (i.e. the sources of most of the food items consumed in the city) (FAO). (The map for Lusake is available at: http://www.fao.org/in-action/food-for-cities-programme/pilotcities/lusaka/en/; the map for Kitwe is available at: http://www.fao.org/in-action/food-for-cities-programme/pilotcities/kitwe/en/.)

Based on the approach and goals defined through each CRFS process, each of the defined city regions has the longer-term aim to make its CRFS more sustainable and resilient, and to improve the livelihoods of rural and urban dwellers in the city region, with special attention to the challenges of: (a) how to improve access to adequate food for the vulnerable and poor urban population; and (b) how to improve market access for the smallholder farmers in urban, peri-urban, and rural areas in the city regions. This connective approach to assessment examined current and future constraints affecting the local and regional food value chain. It used local knowledge to help analyse and prioritize these constraints and explore new ideas to strengthen the sustainability and performance of the food system.

Since very few data were available, an important focus was made on collecting primary data, unlike in the other project pilot cities. Data were collected around the main priorities identified by multi-stakeholder groups to enable solid locally owned strategies and advocacy towards the key local institutions. The multi-stakeholder group was mobilized through a series of workshops to discuss, validate the assessment and identify key strategies to be implemented, and define associated action plans, including timeframe, funding needs and sources, as well as actors to be involved.

The CRFS assessment and planning process played a crucial role in contributing to identifying gaps and bottlenecks to create more resilient and inclusive food systems within specific city regions. In particular, as maize occupies a central position in Zambia's agricultural political economy, the CRFS process highlighted the importance of crop diversification and, specifically, the role of horticultural production and the value chain in feeding the urban population and contributing to healthy nutrition.

As a result of this process, there has been an increased awareness of the importance of joint planning between the two cities and their surrounding districts for the implementation of each CRFS. Joint planning is proposed in the Urban and Regional Planning Act, but guidelines and standards are not available yet. This would provide a policy and institutional framework to anchor implementation processes (FAO-RUAF, n.d.).

In addition, there has been renewed interest and policy discussions at institutional levels. In particular, this process has built bridges of communication among institutions to introduce a more integrated and territorial perspective in planning sustainable food systems. For instance, the CRFS project facilitated dialogue between the Ministry of Agriculture and the Ministry of Local Government on the importance of mainstreaming food and agriculture in the process to decentralize authority from the national level. In each city region, the whole process heightened awareness about the importance of going beyond the sectoral approach when looking at food, and the need to integrate all actors in the discussion.

In the framework of the decentralization process in Zambia, the CRFS assessment and planning process, together with its findings and recommendations, were part of the basis to contribute to the ongoing formulation of the National Urbanization Policy (NUP). The NUP aims to provide an overarching coordinating framework to address urban challenges and to maximize the benefits of urbanization, while mitigating potential adverse externalities. The CRFS assessment and planning process will be essential for providing key inputs to ensure that food security and nutrition, as well as food system dimensions, are part of the policy. In addition, the CRFS process highlighted the challenges that arose as a result of the current fragmented governing bodies for food systems that do not normally work in collaboration; a multi-stakeholder and interinstitutional mechanism or body responsible to define food strategies and policies would be key to reinforce the food system, in order to ensure food and nutrition security, including food safety (Hemmati, 2012; Vervoort et al., 2014). Furthermore, the decentralization policy offers a good platform for setting up a food council as it is linked to certain national government functions such as policy and programme responsibilities for agriculture and health being devolved to local government. This means the local authority, Ministry of Agriculture and Ministry of Health are already working hand in hand but also highlights the other stakeholders that can be included to create well-adjusted strategies and policies, in order to, for example in Lusaka, reinforce the food system.

Case study 2: Colombo, Sri Lanka – from poverty and health focuses to (food) system thinking

Colombo District has more than 2.3 million inhabitants, with a population density of more than 3,300 people/km^2 and increasing. To meet the needs of this growing population, food is sourced from many parts of the country. However, owing to inefficiencies in the wholesale market system, food prices are high, resulting in high levels of food insecurity. There is also concern for food safety, as pesticide use is not well controlled.

Based on a scan of existing institutions and their connectivity as part of the CRFS assessment, it is clear that at the institutional level there is significant fragmentation in regard to food as there is no specific authority in charge of urban food security, agriculture, or rural–urban food supply. As documented in the pilot city synthesis report, there is a considerable number of relevant government departments and authorities at national, provincial, and municipal levels that oversee the system, especially focusing on food supply, prices, and consumer protection (FAO et al., 2018). Potentially adding to the urban–rural divide, the Sri Lankan government has recently set up the Ministry of Megapolis and Western Development (MoMWD), a dedicated ministry to implement Megapolis, a large-scale, multibillion-dollar urban development initiative in Western Province where Colombo is located.

The Colombo city region food system (CRFS) was defined based on: (a) built-up areas and population density (less dense areas of the region could act as suppliers to Colombo city); (b) jurisdictional and administrative boundaries (governing units that take policy decisions); and, (c) supply areas of macro- and micro-nutrients to the Colombo city region (for more information about boundary setting see Blay-Palmer et al., 2018). From a geographical viewpoint, the Colombo city region has been defined as the Colombo Municipal Council (CMC) and district areas. (The map of the Colombo City Region Food System is available at: http://www.fao.org/in-action/food-for-cities-programme/pilotcities/activitiescolombo/en/.)

When the CRFS process was launched, Colombo did not yet have a clear policy commitment and objective to design a more comprehensive and integrated food system agenda involving the rural areas where food is sourced. Nevertheless, at the municipal level, food was already among the priorities of the municipal government in terms of food safety (public health), food waste (waste management), and with attention to food price increases along the value chain (poverty and food insecurity) in the context of wholesale market system inefficiencies (Tacoli, 2006). At the provincial level, food was prioritized through the promotion of urban and peri-urban agriculture strategies and activities. Nevertheless, food was not analysed or planned in a systemic and integrated manner; current policies and programmes on food systems are fragmented and sectoral, while attempts to achieve macro-level improvements are mostly disconnected and in isolation (FAO et al., in press).

Public health, food security, poverty, and waste issues have been the entry points to initiate a comprehensive and integrated assessment of the food system and were instrumental in fostering stakeholder dialogue and building a common understanding on the interdependence of these issues and their relevance in a broad vision of the Colombo city region food system. This eventually determined further political and stakeholder interest and commitment. In that respect, the CRFS process, while targeting the local-level policy context, allowed initiating the policy discussions and processes at local, provincial (regional), and national levels. In addition, it also contributed to increased attention about the importance of integrated regional food system approaches at provincial and national levels. At the municipal level, the CRFS process has helped the Colombo Municipal Council (CMC) and other institutions to understand the Colombo food system in its complexity and has created the basis to build a common vision on a more sustainable and resilient city region food system. The process has allowed identifying the opportunities, challenges, and needs to be addressed. It was indeed instrumental in understanding the importance of how the urban food system and its regional supply chains and flows across the urban–rural spectrum are interlinked with other high-priority policy objectives. As tangible policy outcomes with the potential for a more coherent regional food systems approach, the CMC agreed to introduce local-level by-laws to promote and regulate Reduction, Reuse, and Recycling (RRR) of food waste at the CMC level. As well, they will introduce a separate division dedicated to food safety within the health department and concentrate more on RRR from food waste.

At the regional level, the Western Province population is growing quickly and the regional Government will have the responsibility to ensure appropriate levels of food security and nutrition in a sustainable manner. While this requires long-term, integrated, and holistic food policies and strategies that include all actors involved in the food system, food continues to be excluded in urban and regional planning. Although a territorial approach is mostly beyond the control of local level authorities, the regional level authority (Western provincial council) has started to recognize the importance of the territorial approach in food systems. In addition, the MoMWD has realized the importance of including the CRFS concept and has made a request for support to build on the findings of the CRFS initiative in integrating prioritized areas of food systems into Western Megapolis urban and regional planning (FAO et al., in press).

At the national level it emerged that existing food policies needed to be evaluated and adjusted. This was well taken by the national authorities, and initial policy discussions to integrate the results of the CRFS assessment into the National Agricultural Policy, the National Nutrition Policy, and the Food Act have begun. However, concrete measures to align local, provincial, and national strategies and action plans are still needed.

Overall, the CRFS process has shed light on food safety, nutrient safety, value chain management, and food waste and losses in relation to urban spaces. Using the CRFS framing triggered policy discussions beyond the local level to provincial (regional) and national levels (Blay-Palmer et al., 2018). It has created the basis for starting to visualize the importance of a territorial approach to the food systems and actions needed to offset the impacts of natural resource management, climate change, and shocks on city regions.

Case study 3: Medellin's approach to city region food systems and enhancing rural–urban linkages

The Municipality of Medellin is the second largest city in Colombia after the capital city of Bogota, with close to 2.5 million inhabitants. The department of Antioquia, where Medellin is located, is made up of 125 municipalities grouped into nine sub-regions. The definition of the city region was based on the social, economic, and political dynamics around the food system associated with Medellin and its Metropolitan Area (a region called Valle de Aburrá) that is constituted by ten municipalities, in which 59 per cent of the population of Antioquia resides.

The Medellin city region was defined using five different criteria: (a) food supply: the municipalities contributing more than 1 per cent of the food consumed through the supply centres; (b) production including those municipalities contributing 1 per cent or more of the total food produced in the Department of Antioquia; (c) proximity with those territories within the Aburrá Valley that currently have agricultural production; (d) potential for agricultural expansion based on food-flow analysis by volume (weight in tons); and finally (e) political participation to consider municipalities important to governance even though they do not participate significantly in the production or supply of agricultural products for Medellin (Dubbeling et al., 2017; FAO-RUAF, 2018). (The map for the Medellin City Region Food System is available at: http://www.fao.org/in-action/food-for-cities-programme/pilotcities/medellin/en/.)

The city of Medellin is a pioneer in its approach to urbanization by ensuring the well-being of its citizens through food. For example, food security and nutrition issues have received growing interest from decision makers over the years, which has now spread to the public and media of the city and the Antioquia region. In 2009, Medellin became the first city in Colombia with a dedicated Food and Nutrition Security unit. The role of the municipal authorities in public policy on food and nutritional security has been growing – the government of Antioquia also institutionalized the issue of food security and nutrition as public policy in 2003 through the Management of Food Security and Nutrition (MANA) programme (Hackett et al., 2008).

Accordingly, these food-flow and programme initiatives were the entry points to initiate the CRFS process that started with a solid urban food policy agenda as a basis to promote a city region food system perspective. The process focused on the policy phase to ensure that a CRFS approach was embedded in new political programmes and agendas. Once support was ensured, the CRFS approach continued with the assessment, while work on policy processes proceeded in parallel.

The CRFS assessment and related processes highlighted the strong interdependence of the city with surrounding territories and the need to address issues of fragmentation and the inefficiency of the supply system as well as the significant social and economic territorial inequalities between urban, peri-urban, and rural areas. There are limited direct relationships between rural and urban spaces, with a lack of interaction among producers, marketers, and consumers. As a consequence the food provisioning systems in and around the city of Medellin are quite inefficient, resulting in considerable food loss and waste and limited market regulation for prices, due to the hegemonic role of a limited number of actors involved in food logistics and commercialization that act as an oligopoly without real control on price generation. This points to opportunities to strengthen food production in the peri-urban areas of cities in the Medellin city region, as well as to improve linkages between urban food demand (especially from lower-income neighbourhoods) and cooperatives of small agricultural producers in the city region (Blay-Palmer et al., 2018). Considering interventions in other food systems, public administrations could focus particularly on improving the logistical infrastructure of publicly supported markets and possibly creating 'food hubs' for local food to enter and be distributed throughout the city (Blay-Palmer et al., 2013; O'Connell and Kiparisov, 2018). Local leaders could enable this through inclusive food governance mechanisms that address the complexity of the food system in the city region of Medellin to generate political, administrative, and economic synergies that facilitate the implementation of actions in the city region.

The CRFS process facilitated institutional integration between the regional government of Antioquia, the Medellin Mayor's Office, and the Metropolitan Area of the Valley of Aburrá. Advances have been made in starting the creation of a new governance structure and institutional platform in which the three public authorities collaborate. This tripartite platform on territorial food policy issues, called the 'Alianza por el Buen Vivir' (the 'Alliance for Good Living'), serves as a forum and mechanism for coordination, knowledge sharing, and articulation of the collective development and implementation of policy and project interventions from a territorial governance perspective in the Medellin food system. Some examples of this tri-partite collaboration are: (a) The commissioning of a study to further assess possibilities for the production and commercialization of food products from the region's rural villages; (b) A proposal has been developed to renovate the *Campo Valdes* urban food market into a regional food logistics centre or 'food hub' within the city to make it more accessible for producer associations in the rural areas around Medellin; (c) The platform is also exploring strategies for the regulation of intermediaries to allow these fresh products to reach consumers at much more accessible prices (Dubbeling & Santini, 2018).

Policy outcomes in other city regions

Other city regions participated in the CRFS pilot project, including Toronto, Canada and Quito, Ecuador. In the example of Toronto and the surrounding Greater Golden Horseshoe region, while the CRFS work was only one minor contributor to food policy considerations at multiple scales, pre-CRFS work helped to shape other food policy initiatives either directly or indirectly. In Toronto, complex, long-term, and strong food-related ties between food and governance have existed for many years. For example, the Toronto Food Policy Council was established 30 years ago within the Public Health department of the City of Toronto. This deep history enabled the CRFS process and the coherence of the work by the CRFS Task Force. The multi-stakeholder Task Force included municipal officials from: Toronto Food Policy Council, and Toronto Food Strategy as part of the City of Toronto Public Health as well as City of Toronto Food and Beverage Sector; the provincial Ministry of Agriculture Food and Rural Affairs; a representative from the Golden Horseshoe Food and Farming Alliance; and academic experts in food policy, food security and nutrition, and sustainable food production and food systems. Through the CRFS Task Force review, these key food system actors who normally attended to specific, narrower responsibilities and related policies and programmes were able to come together to identify common issues and cross-cutting solutions to recommend systemic policy changes. The CRFS project contributed to increased awareness about regional food opportunities and concerns, a growing awareness of multi-scaled policy interactions starting with the Toronto Food Policy Council and opportunities to build synergies through, for example, a food–energy–water nexus (Miller and Blay-Palmer, 2018).

In Quito, Ecuador, participatory governance in the territorial food system is recognized as a way to guarantee the achievements of the set goals and outcomes of the food strategy (Dubbeling et al., 2017). These contribute to initiatives across the sustainability spectrum, including the Right to Food, promoting sustainable diets, improving urban–rural linkages and ensuring participation of small producers and local authorities in the city region. The CRFS research in Quito helped to advance this work by supporting network capacity building. Through a consultative process, Quito has designed an appropriate food system governance structure that takes the form of a food policy council, seeking involvement of local, provincial, and national government actors, the private sector, and civil society. This has helped to further reinforce the regional food system and provide a platform for more robust urban/peri-urban/rural networks.

Conclusions

The CRFS approach confirms that each city region food system is unique. Each has its own peculiarities, challenges, and solutions (Marsden, 2013; Sonnino et al., 2016; see also Chapters 2, 3, 4, 6, & 7, this volume). The CRFS process in the different pilot city regions has generated a broad variety of policy results and has faced several challenges that have limited the impacts as documented in the toolkit.

Overall, the assessment process as part of the CRFS pilot projects allowed multiple stakeholders in the city regions to understand how urban and surrounding rural areas are fed and what their food dependencies are and to identify weaknesses and potential pressure points. This raised general awareness and enabled the basis for policy transformation and the implementation of more sustainable and resilient CRFS through targeted strategies to improve their food systems. With the current pressures from climate change and related disasters, the CRFS approach could offer a way to mitigate, adapt to, and prepare for these changes, creating more resilient regions by providing a method to define place-based challenges, identify solutions, and build capacity (Blay-Palmer et al., 2018).

The assessments helped city stakeholders to recognize the interconnections between food and agriculture and several other sectors, such as transport (as a large part of city transport is food-related), health (malnutrition, obesity, school feeding), education (awareness on sustainable diets through curricula), land-use planning for agricultural and food (land allocation for food and green infrastructure, food market relocation), community development and revitalization, employment generation (in food production, processing and retail, food waste management), and waste management (productive use of waste and waste water, management of food waste) (Tacoli, 2006). In addition, a CRFS approach helped cities such as Medellin and Quito to understand the potential and opportunity to shorten the supply and value chains of key foods by localizing production and reinforcing existing local initiatives. The process also helped evaluate the extent to which urban food security is dependent on rural production areas and how the food system impacts both urban and rural populations in the city region. This understanding helps city governments to start seeing food as a driver for other sustainable urbanization policies (Dubbeling & Santini, 2018).

The process also encountered significant challenges and obstacles in operationalizing the CRFS concepts. Some of the hurdles were common in most of the city regions and can be summarized as: (a) limited data availability;

(b) challenges in building political buy-in and stakeholder engagement; and (c) limited governance and regulatory instruments for food planning at the city region level. The CRFS assessment illustrated the significant challenges arising from the scarcity of data and empirical information at the subnational level on food systems. In some cases where data were available, information at the city region level did not match jurisdictional boundaries. In addition, some data were sensitive or subject to copyright. To tackle these challenges, a combination of secondary and primary research, coupled with expert knowledge, was used to complement missing data (Miller & Blay-Palmer, 2018). Addressing data gaps requires identifying innovative and efficient methods to combine secondary information, primary data, and expert opinions and analyse this data in systematic and consistent ways to produce the information required for local decision-making.

As with other approaches, multi-stakeholder dialogue is a crucial element in the process to enable transparent and inclusive participation. Through this, decision makers and organizational representatives have the opportunity to guide implementation and discuss findings and implications for local strategies (Hemmati, 2012). However, any multi-stakeholder dialogue process comes at the cost of a high degree of engagement across as many sectors and stakeholders as possible. Engagement requires time and resources that can be challenging for many stakeholders to commit. In many cases, engaging with key stakeholders may be difficult due to other reasons – for example, lack of institutional versus individual engagement, conflicting agendas, no history of collaboration, and/or no clear outputs from the start of the process. The identification of a political champion (a recognized and respected policymaker from a key institution involved in the CRFS) from the very beginning of the process can be a successful driver to ensure the engagement of key stakeholders and political buy-in (Bagdonis et al., 2009; Kania & Kramer, 2015). The involvement of key stakeholders throughout the process can also ensure ownership and commitment as the policy or action plan will be shaped – as much as possible – according to the needs, demands, and contributions of all the stakeholders involved (Vervoort et al., 2014).

In most of the city regions involved in the pilot programme, appropriate governance structures and regulatory instruments often do not exist that allow for multidimensional and multi-sector food systems planning or facilitate the realization of policies and investments to reinforce the CRFS. In most cases, food policy, if it does exist, is segmented by particular areas of interest, for example public health or farming, and does not have a strong cross-sectoral mandate (Jenning et al., 2015). As revealed in the CRFS pilot process, and consistent with other multi-stakeholder initiatives (Rivera-Lirio & Muñoz-Torres, 2010; de Zeeuw & Dreschel, 2015), governance arrangements are the key to promoting and operationalizing the CRFS concepts – putting the right structures in place to drive and facilitate the creation of new kinds of rural–urban linkages. Accordingly, a critical challenge is creating more inclusive territorial governance structures in which cities, regions, and other levels of government can work constructively together towards complementary, beneficial outcomes (Jenning et al., 2015). Interaction, coordination, and joint planning are necessary between different institutions and levels of governments involved in the CRFS (urban and rural entities, larger and smaller cities in the city region, city and provincial/national government). While there is value in integrating across different sites of food production to include rural, peri-urban, and urban agriculture in a more linked up manner (Neilson & Rickards, 2017; Clancy & Ruhf, 2018) typically, such institutions, urban and rural authorities, or city level versus provincial authorities, do not often have the institutional capacity for engaging in joint policy and planning, due to the limitation of their jurisdictional mandate or when different political orientations are at play. As elaborated in the toolkit, task forces and institutional focal points were key enablers to realizing improved linkages and more effective policy and programmes. This, in turn, can foster more sustainable, resilient food systems. This applies to all city region food systems – ones that exist, are being revitalized or are in their beginnings. Building resilient and sustainable CRFS requires opening space for democratic participation from all spheres of society, fostering a multi-stakeholder dialogue process so that citizens can play a stronger role in the policy development process. In future work, it will be essential to include medium and smaller cities and their regions as this is where the most people in the world live and so would have the most potential for impact (Berdegué et al., 2015) and to broaden the scope of the toolkit to include considerations such as climate resilience and migration issues. Using the toolkit's CRFS approach, five pilot communities were able to gain new insights, but also discover important limitations. The toolkit provides a suite of tools from visioning, assessment, and policy creation so that city regions can activate to enable sustainable food system change.

Acknowledgements

We are grateful for the financial support of the German Federal Ministry of Food and Agriculture, the Daniel and Nina Carasso Foundation, and the CGIAR Water Land and Ecosystems Programme led by IWMI, and the Social Sciences and Humanities Research Council of Canada that enabled this research.

References

Bagdonis, J.M., Hinrichs, C.C., & Schafft, K.A. (2009) The emergence and framing of farm-to-school initiatives: Civic engagement, health and local agriculture. *Agriculture and Human Values*. 26(1–2), pp. 107–119.

Berdegué, J.A., Carriazo, F., Jara, B., Modrego, F., & Soloaga, I. (2015) Cities, territories, and inclusive growth: Unraveling urban–rural linkages in Chile, Colombia, and Mexico. *World Development*. 73, pp. 56–71.

Biriwasha, M. (2017) *HIVOS: Opinion. Why media matters to sustainable food in Africa*. Report available at: https://southern-africa.hivos.org/opinion/why-media-matters-to-sustainable-food-in-zambia/.

Blay-Palmer, A., Landman, K., Knezevic, I., & Hayhurst, R. (2013) Constructing resilient, transformative communities through sustainable 'food hubs'. *Local Environment*. 18(5), pp. 521–528.

Blay-Palmer, A., Renting, H., & Dubbeling, M., (2015). City Region Food Systems: A literature review. *Carasso Foundation*. Available at: https://www.ruaf.org/sites/default/files/City%20Region%20Food%20Systems%20literature%20review.pdf.

Blay-Palmer, A., Santini, G., Dubbeling, M., Renting, H., Taguchi, M., & Giordano, T. (2018) Validating the city region food system approach: Enacting inclusive, transformational city region food systems. *Sustainability*. 10(5), pp. 1–23. Available at: https://doi.org/10.3390/su10051680.

Clancy, K. & Ruhf, K.Z. (2018) New thinking on 'regional'. *Journal of Agriculture, Food Systems, and Community Development*. Advance online publication. Available at: https://doi.org/10.5304/jafscd.2018.083.008.

de Zeeuw, H. & Drechsel, P. (Eds.) (2015) *Cities and agriculture: Developing resilient urban food systems*. London, Routledge.

Dubbeling, M. & Santini, G. (2018) City region food systems assessment and planning. *RUAF Urban Agriculture Magazine*. May(34), pp. 6–9. Available at: https://www.ruaf.org/sites/default/files/RUAF_UAM34_p6-9.pdf.

Dubbeling, M., Santini, G., Renting, H., Taguchi, M., Lançon, L., Zuluaga, J., De Paoli, L., Rodriguez, A., & Andino, V. (2017) Assessing and planning sustainable city region food systems: Insights from two Latin American cities. *Sustainability*. 9(8), p. 1455. Available at: https://doi.org/10.3390/su9081455.

FAO, IWMI & RUAF (in press) *Assessing and planning city region food system, Colombo (Sri Lanka) Synthesis Report*. Rome, FAO.

FAO & RUAF (2015) *A vision for City Region Food Systems*. Available at: http://www.fao.org/3/a-i4789e.pdf.

FAO & RUAF. (n.d.) Building more sustainable and resilient food system in the Lusaka city region: Policy brief. Available at: http://www.fao.org/3/I9954EN/i9954en.pdf.

FAO, RUAF & Laurier Center for Sustainable Food Systems (LCSFS) (2018) *CRFS toolkit*. Available at: http://www.fao.org/in-action/food-for-cities-programme/toolkit/introduction/en/.

Government of Zambia. 2015. Urban and Regional Planning Act No. 3 0f 2015. Available at: http://www.parliament.gov.zm/sites/default/files/documents/acts/The%20Urban%20and%20Regional%20Planning%20%20Act%2C%202015.pdf.

Hackett, M., Melgar-Quinonez, H., & Uribe, M.C.A. (2008) Internal validity of a household food security scale is consistent among diverse populations participating in a food supplement program in Colombia. *BMC Public Health*. 8(1), p. 175.

Hemmati, M. (2012) *Multi-stakeholder processes for governance and sustainability: beyond deadlock and conflict*. London, Routledge.

Jenning, S., Cottee, J., Curtis, T., & Miller, S. (2015) *Food in an urbanised world: The role of city region food systems in resilience and sustainable development*. London, 3Keel.

Kania, J. & Kramer, M. (2015) The equity imperative in collective impact. *Stanford Social Innovation Review*. October, pp. 36–41.

Lusaka Central Statistical Office (2010) *Zambia, Lusaka province analytical report: 2010 census of population and housing*. Available at: https://www.zamstats.gov.zm/phocadownload/2010_Census/2010_Census_Analytical_Reports/Lusaka%20Province%20Analytical%20Report%20-%202010%20Census.pdf.

Lusaka Government (2015) *Lusaka urban and regional planning act 2015*. Available at: http://www.parliament.gov.zm/sites/default/files/documents/acts/The%20Urban%20and%20Regional%20Planning%20%20Act%2C%20 2015.pdf.

Marsden, T. (2013) Sustainable place-making for sustainability science: The contested case of agri-food and urban–rural relations. *Sustainability Science*. 8(2), pp. 213–226.

Miller, S. & Blay-Palmer, A. (2018) *Assessment and planning of the Toronto City Region Food System – Synthesis Report*. The Hague, Prepared for the RUAF Foundation.

Neilson, C. & Rickards, L. (2017) The relational character of urban agriculture: Competing perspectives on land, food, people, agriculture and the city. *The Geographical Journal*. 183(3), pp. 295–306.

O'Connell, J. & Kiparisov, P. (2018). Kyrgyzstan value chain gap analysis. Rome, FAO. http://www.fao.org/3/i9199en/I9199EN.pdf

Rivera-Lirio, J.M. & Muñoz-Torres, M.J. (2010) Sustainable development in the Spanish region of Valencia and the social responsibility of SMEs: A multi-stakeholder vision on the role of public administrations. *Journal of Environmental Planning and Management*. 53(5), pp. 573–590. doi:10.1080/09640561003694419.

Sonnino, R., Marsden, T., & Moragues-Faus, A. (2016) Relationalities and convergences in food security narratives: Towards a place-based approach. *Transactions of the Institute of British Geographers*. 41(4), pp. 477–489.

Tacoli, C. (Ed.) (2006) *The Earthscan reader in rural–urban linkages*. International Institute for Environment and Development. London, Routledge.

Vervoort, J.M., Thornton, P.K., Kristjanson, P., Förch, W., Ericksen, P.J., Kok, K., Ingram, J.S., Herrero, M., Palazzo, A., Helfgott, A.E., & Wilkinson, A. (2014) Challenges to scenario-guided adaptive action on food security under climate change. *Global Environmental Change*. 28, pp. 383–394.

World Bank (2016) World Bank micro data catalogue: urban population (% of total). Paris, World Bank. Available at: https://data.worldbank.org/indicator/SP.URB.TOTL.IN.ZS.

10 Assessing responsible food consumption in three Ecuadorian city regions

Myriam Paredes, Donald C. Cole, Fabian Muñoz, Gabriel April-Lalonde, Yubari Valero, Priscila Prado Beltrán, Laura Boada, Peter R. Berti, and the Ekomer project team

Introduction

Most examinations of sustainability of agrifood flows in Latin America have tended to focus on production, distribution, and consumption as separate processes (Gustafson et al., 2016). Such a disarticulated view inhibits the understanding of complex phenomena related to agrifood systems, such as the nutritional transition and the epidemic of obesity in Latin American countries. A more holistic vision of food consumption requires identifying patterns in which the origin of foods, demands of food consumers, and the ways in which provisioning occur are all woven together (Blay-Palmer, 2006; Sonnino, 2009; Csutora & Vetöné, 2014; see also Chapters 3, 4, 6, 9, & 11, this volume).

Social, economic, and political pressures reinforce industrialized food systems (Dixon, 2009; Guthman, 2014). On the other hand, food environment characteristics, such as the availability of agroecological spaces, can influence food consumption practices (Brug, 2008). Further, consumers are not homogeneous, either because of limitations in resources or because of active choices. Some individuals, households, and organized groups search out lifestyles and provisioning approaches which are more sustainable, based on differing values and meanings (Stern & Dietz, 1994; Ozçaglar-Toulouse, 2009; Inglehart, 2015). They go beyond simple market transactions towards concerns about health (Gould, 1988; Moorman & Matulich, 1993), ecology and environment (Kinnear et al., 1974; Zimmer et al., 1994), corporate social responsibility (Ottman & Reilly, 1998; Pivato et al., 2008; Feldman & Reficco, 2015; Tsai et al., 2015; Frynas & Yamahaki, 2016), ethics (Newholm & Shaw, 2007; Linders, 2014), and individual social responsibility (Berkowitz & Lutterman, 1968; Anderson & Cunningham, 1972). They engage in what we denote here as 'responsible consumption' (Webster, 1975; Antil, 1984; Dueñas Ocampo et al., 2014).

Although qualitative approaches have been used to understand the motivations of individuals or household members in making decisions to be responsible consumers (see, for instance, Guerrón-Montero & Moreno-Black, 2001; Piñeiro & Díaz, 2012; López et al., 2017), limited quantitative research has characterized the extent to which entire populations engage in responsible consumption. Key questions remain: how can one define and measure responsible food consumption among consumers in city region agri-food systems? (see also Chapters 2 and 9, this volume.), and what relationships might consumption have with healthy eating practices? (see Chapter 7, this volume).

This chapter starts with the context in which we worked, the conceptualization of dimensions of responsible food consumption, and the empirical approach that we took to assessment. We then share our initial results of measurement of the dimensions and the overall Responsible Consumption Index (RCI) and its relation to healthy eating indicators. We discuss the implications of our work and conclude with potential directions for research and application.

Context

In Ecuador, 62 per cent of the population between 19 and 60 years of age are overweight or obese (Freire et al., 2014). Closely related is the burden of chronic diseases (GBD, 2017), which in Ecuador have been estimated to cost society €1.5 billion annually (MIES et al., 2017), approximately 1.5 per cent of GDP. While these chronic diseases are often denoted as non-communicable, they can nevertheless be socially transmitted conditions, being shared among populations and fostered by industrialized agrifood systems that promote highly processed foods (Allen & Feigl, 2017).

Ecuador has great potential for resolving both obesity and chronic diseases, as the vast majority of foods consumed are produced in the country, and 60 per cent of these are produced on diversified, family farms (MAGAP, 2016). However, market chains usually involve multiple intermediaries, leaving farm families with insufficient recompense for their production efforts, pushing them to expand the production area devoted to more marketable crops and reduce the area for vegetables, fruits, pulses, and oilseeds for home consumption (Carrión & Herrera, 2012). Further, diets are predominantly composed of tubers and grains, with low intakes of vegetables and fruits (Freire et al., 2013). In 2008, the government of Ecuador included food sovereignty in its constitution, as one way to reduce or eliminate undernutrition and malnutrition. Its mandate was to promote nutritious food, with a preference for agroecological and organic production, which comes from micro-, small-, and middle-sized peasant production and artisanal fisheries, as well as to foster popular economic organizations (Government of Ecuador, 2010).

The Ekomer research team, a multidisciplinary team of Ecuadorian and international universities and civil society organizations, arose out of both a concern for chronic diseases and a recognition of the potential that Ecuador offers for addressing this problem. The team has carried out research to understand the conditions in which social movement campaigns promote responsible consumption and public policies that support it. One challenge was to develop and implement a method to elucidate different 'responsible (food) consumption' patterns in three counties (*cantones*) centred in city regions (including urban and rural areas) where a citizen campaign for responsible consumption has been particularly active: Ibarra, Quito, and Riobamba in the provinces of Imbabura, Pichincha, and Chimborazo respectively, all in the central Sierra of Ecuador.

Quito's population of 2.2 million is ten times that of Riobamba and twelve times that of Ibarra (INEC, 2010). Of the three provinces, the prevalence of undernutrition, as assessed by stunting (low height for age in 0–5-year-olds), is greatest in Chimborazo (49 per cent), followed by Imbabura (35 per cent) and Pichincha (29 per cent) – national average is 25.3 per cent (ENSANUT-ECU, 2012). On the other hand, overweight ($25 \leq$ BMI[1] < 30) and obesity (BMI ≥ 30) prevalence among adults 20–59 years of age is highest in Imbabura (62 per cent), followed by Pichincha (55 per cent) and Chimborazo (53 per cent), whereas the national average is 62.8 per cent (ENSANUT-ECU, 2012). Among adolescents, a study that included Quito found that the condition of being overweight was associated with inactivity due to >28 hours weekly watching television and high consumption of processed foods (Yépez et al., 2008).

In a national survey of household incomes and expenses (ENIGHUR-INEC, 2013), neighbourhood stores were the most common location for food purchases but the least common location for purchase of less processed foods. In contrast, fairs (open air markets), where the greatest purchase of non- or minimally processed foods occurred (Muzo et al., 2017, p. 28), were the second most common location for food purchases. At the same time, in the three study counties, the majority of agricultural production is dedicated to consumption within the country, rather than export (INEC-ESPAC, 2017), opening up the possibility of greater self-sufficiency in food production in the food-sheds of the selected counties.

Dimensions of responsible consumption

Dueñas Ocampo and colleagues (2014) reviewed the history of socially responsible consumption as a concept, from a personal psychological attribute to a collective behaviour that encompasses environmental, ethical, and social concerns linked with purchasing considerations beyond price. They defined a socially responsible consumer as 'one who sees in their consumption the opportunity to conserve the environment and the quality of life in society in a particular, local context' (p. 289). They noted that most studies are strongly influenced by an economic perspective, centred on demand and terms of exchange in the purchasing of products and services.

In taking a more social perspective on consumption, we have adapted our responsible food consumption concept from Alan Warde's (2005) definition of consumption. Thus, responsible food consumers consciously appreciate and appropriate patterns of production, distribution, use, and recycling of food goods and services, which they render more sustainable. Such responsible consumers are interested in knowing where food comes from, the way in which it was produced, the working relationships involved, and the political and environmental implications of their form of consumption in society at large (Antil, 1984; Newholm & Shaw, 2007; Dueñas et al., 2014). Here, we add the efforts of consumers to self-organize around ethical values and morals of consumption, and to exert political influence at any stage of the process. Such is the notion of 'co-producer', 'a consumer who knows and understands problems of food production' (Carlos Petrini in Beccaria, 2016).

Focusing on responsible consumption in food systems, Heinisch (2016) emphasizes the importance of considering sustainability in the entire set of relationships across the life cycle of food. A food system consists of all the elements (environment, people, inputs, processes, infrastructures, institutions, etc.) and activities that relate to the production, processing, distribution, preparation and consumption of food, disposal of food waste, and the outcomes of these activities, namely nutrition and health status, socio-economic growth, and equity and environmental sustainability (HLPE, 2014). Research on responsible food consumption is scarce, but 'responsible' is often used interchangeably with 'sustainable' when studying food consumption from this perspective. Sustainable diets, as they have been defined and studied, mainly explore the relationships between eating behaviours, health, and environmental impact indicators (Mertens et al., 2016). In normative terms, sustainable diets are protective and respectful of biodiversity and ecosystems; culturally acceptable and accessible; economically fair and affordable; nutritionally adequate, safe, and healthy; while optimizing use of natural and human resources (Burlingame & Dernini, 2012). Agroecological production refers to limited use of external inputs, natural resources conservation, equity and social justice, limited geographic distances (local), appropriate to seasonal availability, and healthy for people and ecosystems (Blay-Palmer, 2006; FAO, 2010, 2018; Lang & Heasman, 2015). Responsible consumption should be oriented towards the broader goal of satisfying the food needs of the entire population of a region in an equitable way, one which maintains the ecological integrity of agroecosystems and the health of the population (Fraňková & Haas, 2017).

The complexity of the concept of responsible consumption has meant that different researchers have included different dimensions in quantitative instruments. In practice, any one approach to measurement cannot capture all relevant aspects of responsible consumption (Lecompte, 2005), rather there should be efforts to adapt them to particular contexts and needs. In the geographic and cultural context of Ecuador, and according to exploratory ethnographic studies with families in Quito (Maas, 2017), we decided to include three dimensions:

1) Direct purchase from producers, as an indicator of contribution to the local community and to smallholder farmers' economy.
2) Preference for agroecological products, as an indicator of a preference for more sustainable ways of food production.
3) Consumption of Andean grains, as an indicator of appreciation of local gastronomic culture.

Dimension One: direct purchase from producers

For the direct purchase from producers, we consider the locations and forms of procuring foods. As smallholder farmers are the most common type of farmers in Ecuador and they primarily produce diverse products for national consumption (MAGAP, 2016), responsible food consumption must consider the sustainability of their livelihoods. Unfortunately, large chains of intermediaries impact smallholders' livelihoods (Chauveau & Taipe, 2012), hence direct purchase from producers demonstrates a sense of co-responsibility for smallholders' well-being among consumers. Face-to-face meeting of producers and consumers generates a greater sense of solidarity, based on fair prices, increasing the incomes of smallholder producers. Acquiring foods directly promotes virtuous spirals of relationships that have been well documented as short circuits of food commercialization (González et al., 2012; CEPAL, 2014; Craviotti & Soleno Wilches, 2015; Contreras et al., 2018), alternative circuits of commercialization (Chauveau & Taipe, 2012), or local agrifood systems (Cerdán, 2014). Hence, various forms of consumer food procurement can bolster community economies: through direct purchase from farmers, at farmers' markets or fairs; through food baskets, as in community supported agriculture; or via meals in restaurants which buy directly from smallholder producers.

Dimension Two: preference for agroecological products

Dimension Two reflects concern about the way foods are produced. In Ecuador, agrarian reform and agricultural modernization efforts in the 1970s undermined existing knowledge and diverse production practices through the intense promotion of mechanization and synthetic agrochemical inputs and fewer crop varieties (Suquilanda, 2006) leading to erosion, declines in soil quality, and adverse human health impacts (Cole et al., 2007; Sherwood, 2009; Paredes, 2010). To address these challenges, more ecological production practices have been promoted (Fundación Heifer, 2014), including agroecology. Agroecology is based on ecological principles such as the conservation of spatial and temporal biodiversity, sustainable management of soils, recycling of nutrients, use of sustainable energy inputs, and biological control of pest populations (Altieri, 1999; Gliessman, 2007; Sarandón & Flores, 2014). Purchase of agroecological foods fosters both sustainable agrifood systems and environmental balance (FAO, 2018). Hence, consumer purchases of agroecological products is valued in this dimension.

Dimension Three: consumption of Andean grains

Dimension Three is represented by an indicator of consumption of three highly nutritious Andean grains: quinoa (*Chenopodium quinoa*), amaranth (*Amaranthus caudatus*), and chocho (*Lupinus mutabilis*). These three crops have been cultivated for millennia in the Andes with several varieties adapted to zones with depleted soils and limited water availability (Peralta et al., 2012). Quinoa and amaranth have higher protein and lower carbohydrate content than grains such as rice and wheat, which have become more common in the Ecuadorian diet (Jacobsen & Sherwood, 2002; Freire et al., 2013). Chocho provides essential fatty acids, approximately 22 per cent by dry weight (Villacrés et al., 2010). Hence, all three grains can contribute to healthy diets and address both under- and over-nutrition in the Ecuadorian population, support production by smallholder farm families, avoid their disappearance in local production, and promote cultural heritage and traditional cuisine (Unigarro Solarte, 2010; Ministerio de Cultura y Patrimonio, 2013).

Empirical approach

Questionnaire design and surveying

We designed a household questionnaire to capture the different dimensions of responsible consumption described above. Exploratory ethnographic work provided an opportunity to adapt the questions to the understandings and context of households in Quito (Maas, 2017; see also Chapter 7, this volume). The questionnaire as a whole consisted of 78 questions, which also addressed topics other than responsible consumption: ten questions about general household characteristics, 22 questions about household food acquisition practices, and 36 questions about individual dietary practices and knowledge. Interviewers were trained by the lead authors in two-day workshops, followed by one day of practice interviews. The training included how to select the respondents within the selected households, how to ask each question, and how to record the data on Android tablets. For all data collection, interviewers used Android tablets with a pre-coded interview guide that was constructed using ODK (https://opendatakit.org/). The latter obviates a separate data entry step and permits daily monitoring of incoming data as soon as data are uploaded to a cloud-based server.

In each of the three study counties, a two-staged, random sample of households was selected to represent both urban (64–74 per cent) and rural (26–36 per cent) populations. First, census sectors, subdivisions of counties defined by the Ecuadorian National Institute of Statistics and Censuses (INEC), were selected randomly. Within each *manzana* (roughly translates to 'neighbourhood') of the chosen census sector, ten dwellings were chosen randomly. As necessary in multi-household dwellings, one household was chosen randomly within that dwelling. Within each household, we explained the project objectives, sought written consent (authorized by the Bioethics Committee of the San Francisco University of Quito), and interviewed two people: a principal adult respondent who answered questions on food provision in the household, and a second adult respondent of the opposite sex. When there was more than one eligible principal or second adult, we randomized by selecting the one with the most recent birthday. Response proportions were high: Ibarra (1282/1475, 87 per cent), Quito (775/860, 90 per cent), Riobamba (858/896, 96 per cent).

For surveys in agroecological locations, the same team of interviewers visited agroecological fairs, markets, stores, and food basket distribution points. Interviewers approached shoppers as they were exiting after their purchases. They explained the study and, when consent was obtained, conducted the interview immediately, except for a few cases where arrangements were made to visit the shopper later in their homes. After the completion of a survey, the interviewers would repeat the process, approaching the next shopper who had completed shopping. The number of agroecological locations was greater in Quito (37) and Riobamba (11) than in Ibarra (6), resulting in larger numbers of respondents in the first two counties (551, 299, and 48 respectively).

Variable and index construction

For each variable, more points are indicative of responses more positively reflective of that dimension (see Table 10.1).

For each household, the scores for each dimension's variable were reduced to the same range of 0 to 3. The variables that make up the dimensions are on an ordinal scale, going from the absence of the characteristic to a significant presence (for example, zero consumption of Andean grains, to at least seven times per month). The determination of scales from 0 to 3 was carried out through a validation in an expert consultation (consensual validity) (Kaplan et al., 1976; Roberts et al., 2006). Developing a common four-point scale across the three dimensions allowed us to standardize the value of the indicators and compare the dimensions for each population subgroup directly (as recommended by Sarandón & Flores, 2014).

The three dimensions were combined into the RCI with different weightings. Our assignation of weights was guided by both the Ecuadorian Andean context and prevalence observed in our surveyed population. As

Table 10.1 Component dimensions of the Responsible Consumption Index (RCI): variables and scoring system

Dimension	*Variable*	*Responses considered part of Responsible Consumption*	*Scoring system*	
			Options	*Score*
1. Direct purchase from producers	Places of food procurement	Direct purchase from producer, purchase at farmers' market, fair, or food basket, grows own, or purchases at agroecological restaurant	If these procurement options are:	
			Important	3
			moderate	2
			minor	1
			rarely or never	0
2. Preference for agroecological produce	Production approach for foods procured	Procured foods produced using agroecological methods	If the procured foods were produced using agroecology methods:	
			Important	3
			moderate	2
			minor	1
			rarely or never	0
3. Consumption of Andean grains	Consumption of Andean grains	Consumption of each of quinoa, amaranth, and chocho	Times per month	
			≥7	3
			5–6	2
			2–4	1
			≤1	0

smallholder family farms provide the majority of national production, but links between urban and rural areas need strengthening in order to promote both food security and access to healthier foods, we allocated the greatest weight to Dimension One (42 per cent). Given the impact of agroecological food production on the environment, as well as on human health, we assigned the second largest weight to Dimension Two (33 per cent). Only 19 per cent of respondents consumed Andean grains more than three times per month, so we assigned a lower weight of 25 per cent for Dimension Three. A household's RCI was then calculated as:

RCI = Dimension 1 score x 0.42 + Dimension 2 score x 0.33 + Dimension 3 score x 0.25

Given skewed distributions, rank correlations were calculated among dimensions and the RCI.

RCI relationships with nutrition relevant practice indicators

Fruit and vegetable consumption was dichotomized into daily versus not. For active control of table salt in the diet, a score was constructed by allocating one point to each control strategy among: (1) minimize consumption of processed foods; (2) and (3) examine food labels for table salt; (4) do not add salt at the table; (5) and (6) buy foods low in table salt; (7) and (8) add little/no salt when cooking; (9) use other spices instead of salt when cooking; and (10) avoid eating away from home. Summed, the salt control practices score could range between 0 and 10.

Comparisons of measures of central tendency of the RCI and of these nutrition-relevant practice indicators were carried out across samples and counties using non-parametric tests of inference: Wilcoxon-Mann-Whitney test for the dichotomous fruit and vegetable consumption; and Spearman correlations for RCI and salt control.

Results

Dimension distribution across different counties and samples

The descriptive statistics on the dimensions are set out in Table 10.2. Direct purchase from producers (Dimension One) and preference for agroecological produce (Dimension Two) have significantly higher scores for consumers from agroecological fairs than in the general population. For both dimensions, Quito has the highest score among consumers of agroecological fairs, while Riobamba has the highest score among consumers of the general population.

For each dimension and for RCI: lower case superscripts with same letter indicate equivalence across the three counties, within the same sample type (random sample of households and fair sample); UPPER CASE superscripts with same letter indicate equivalence across sample type (random sample of households and fair sample), within the same canton.

For consumption of Andean grains (Dimension Three), the population attending agroecological fairs also had a higher average consumption of traditional Andean foods than the randomly selected general population. Quito had the highest score among consumers of agroecological fairs, while Ibarra had the highest score among consumers of the general population.

Table 10.3 presents the rank correlations between the RCI and the three dimensions that make up the index, differentiated by sample. All correlations are significant ($p<.0001$) with the exception of Dimension One (D1) versus Dimension Three (D3) in the fairs sample ($p=0.51$). This is expected since Dimension Two (D2), purchases in agroecological fairs, is usually linked to direct purchasing from farmers (D1), while Andean grain consumption (D3) is not necessarily conditioned on direct or agroecological purchases.

Table 10.2 Descriptive statistics of three dimensions* and overall RCI, by sample type and county

Dimension		*Random*			*Agroecological Fairs*		
		Ibarra (n=1284)	*Quito (n=769)*	*Riobamba (n=861)*	*Ibarra (n=48)*	*Quito (n=551)*	*Riobamba (n=299)*
1– Direct purchase from producers	Mean (SD)	1.26 (0.17)^{aA}	1.08 (0.19)^{bA}	1.27 (0.28)^{cA}	1.42 (0.21)^{aB}	1.52 (0.29)^{bB}	1.42 (0.20)^{aB}
	Median	1.24	1.06	1.29	1.47	1.53	1.41
	(Min–Max)	(0.65–1.76)	(0.47–2.06)	(0.35–2.18)	(1.06–2.12)	(0.82–2.53)	(0.76–2.00)
2 – Preference for agroecological produce	Mean (SD)	0.13 (0.46)^{aA}	0.18 (0.60)^{aA}	0.30 (0.81)^{bA}	1.81 (1.02)^{aB}	2.34 (0.86)^{bB}	2.24 (0.92)^{bB}
	Median	0.00	0.00	0.00	2.00	3.00	2.00
	(Min–Max)	(0.00–3.00)	(0.00–3.00)	(0.00–3.00)	(0.00–3.00)	(0.00–3.00)	(0.00–3.00)
3 – Consumption of Andean grains	Mean (SD)	0.70 (0.50)^{aA}	0.55 (0.45)^{bA}	0.47 (0.29)^{cA}	0.86 (0.57)^{aB}	1.15 (0.73)^{bB}	0.57 (0.28)^{cB}
	Median	0.75	0.50	0.50	0.75	1.00	0.50
	(Min–Max)	(0.00–2.75)	(0.00–2.75)	(0.00–2.00)	(0.25–3.00)	(0.00–3.00)	(0.00–2.25)
Responsible Consumption Index (RCI)	Mean (SD)	0.75 (0.22)^{aA}	0.65 (0.28)^{bA}	0.75 (0.33)^{cA}	1.41 (0.42)^{aB}	1.70 (0.46)^{bB}	1.48 (0.33)^{aB}
	Median	0.71	0.58	0.67	1.41	1.78	1.56
	(Min-Max)	(0.37–2.03)	(0.25–2.01)	(0.17–1.94)	(0.59–2.38)	(0.43–2.53)	(0.51–2.12)

*standardized across different dimensions to a range of 0 (low) to 3 (high), but not according to sampling weights.

Table 10.3 The Rank correlation between the RCI and the three dimensions that make up the index

Spearman correlation coefficients				
	RCI	*D1*	*D2*	*D3*
RCI		*0.67*	*0.80*	*0.59*
D1 (Direct purchase)	*0.61*		*0.44*	*0.16*
D2 (Agroecological preference)	*0.52*	*0.15*		*0.25*
D3 (Andean grains)	*0.63*	*0.01*	*0.075*	

Note: The values above the diagonal are for the random household sample (n=2914). The values below the diagonal are for the sample recruited at agroecological fairs (n=898). All correlations are significant (p<.0001) with the exception of D1 vs D3 in the fairs sample (p=0.51).

Given the correlations, a similar pattern was observed, with distributions in all three cities further to the right (higher) among those attending agroecological fairs than the general population (see Figure 10.1).

RCI relationships with nutrition relevant practices

Marked heterogeneity was observed in fruit and vegetable consumption across counties and samples (see column cell percentages in Table 10.4). Those buying food at agroecological fairs on average were more likely to report daily consumption of fruits and vegetables compared to the randomly selected population (higher column percentages in lower row). Although generally those households reporting daily fruit and vegetable consumption had higher RCI scores, exceptions occurred (e.g. Ibarra agroecological fairs' sample). (Within each column † versus ‡ indicates different values across Fruit & Vegetable strata (No vs Yes, within columns). Within each Fruit & Vegetable strata (No row and Yes row): lower case superscripts with same letter indicate equivalence across the three counties, within the same sample type (random sample of households and fair sample); UPPER CASE superscripts with same letter indicate equivalence across sample type (random sample of households and fair sample), within the same canton. All comparisons were made with the non-parametric Wilcoxon-Mann-Whitney test.)

All counties and samples presented very low mean and median scores (less than 1 out of 10) on active control of table salt in the diet. While there were some significant differences (see Table 10.5), the scores were very low in both random and fair samples. The correlation between regular active control over table salt with the RCI scores was also low (from 0.07 to 0.19). The scores for actively control table salt in the diet are between 0 to 10 according to control strategies: (1) minimize consumption of processed

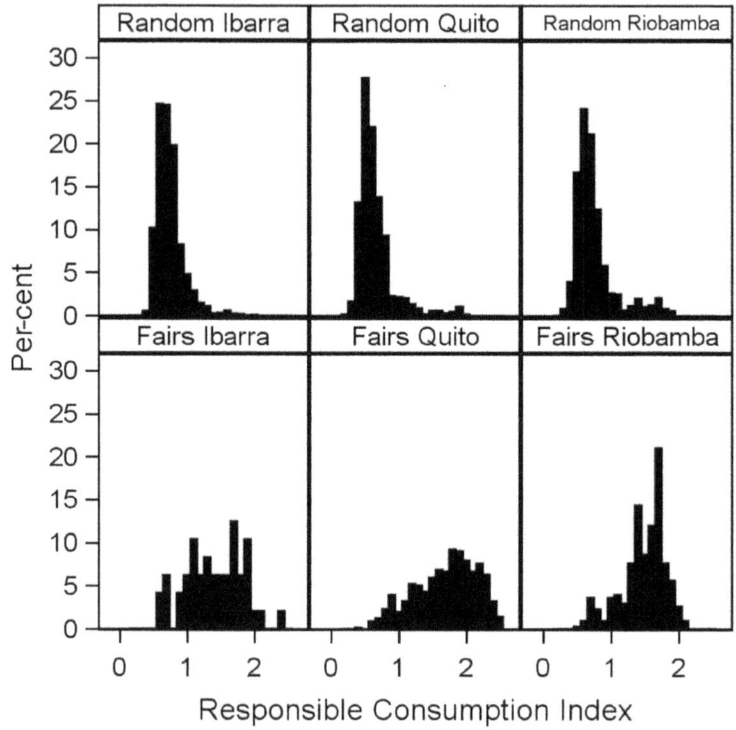

Figure 10.1 Distributions* of RCI by sample type (rows) and county (columns).
* Per cent of sample used to take into account weighting for random samples of households.

foods, (2) and (3) examine food labels for salt/sodium, (4) do not add salt at the table, (5) and (6) buy foods low in salt/sodium, (7) and (8) add little/no salt when cooking, (9) use other spices instead of salt when cooking, (10) avoid eating away from home. (Within the same sample type (random sample of households or fair sample) lower case superscripts with same letter indicate equivalence across the three counties; UPPER CASE superscripts with same letter indicate equivalence across sample type (random sample of households and fair sample), within the same county. All comparisons were made with the non-parametric Wilcoxon-Mann-Whitney test.)

Discussion and implications

Overall, the results show that food environments of each county have influenced engagement in responsible consumption and decisions on healthy

Table 10.4 RCI distributions* by fruit and vegetable consumption category, sample and county

Fruit & Vegetables Daily		*Random*			*Agroecological Fairs*		
		Ibarra (n=1284)	*Quito (n=769)*	*Riobamba (n=861)*	*Ibarra (n=48)*	*Quito (n=551)*	*Riobamba (n=299)*
No	Mean (SD)	0.72 (0.20)^{aA†}	0.61 (0.24)^{bA†}	0.73 (0.32)^{cA†}	1.54 (0.38)^{aB†}	1.62 (0.47)^{aB†}	1.48 (0.31)^{abB†}
	Median	0.68	0.57	0.66	1.61	1.71	1.56
	(Min-Max)	(0.37–1.91)	(0.25–1.92)	(0.17–1.94)	(0.74–2.07)	(0.66–2.43)	(0.51–2.08)
	% of column	56.5	46.7	83.8	37.5	22.1	80.6
Yes	Mean (SD)	0.79 (0.29)^{aA‡}	0.69 (0.30)^{bA‡}	0.87 (0.63)^{aA‡}	1.34 (0.43)^{aB†}	1.72 (0.45)^{bB‡}	1.49 (0.39)^{aB†}
	Median	0.74	0.61	0.78	1.25	1.80	1.56
	(Min-Max)	(0.40–2.03)	(0.25–2.01)	(0.42–1.85)	(0.59–2.38)	(0.43–2.53)	(0.71–2.12)
	% of column	43.5	53.3	16.2	62.5	77.9	19.4

* adjusted according to sampling weights.

Table 10.5 Summary statistics* of table salt control practice scores by sample and county, and correlations between table salt scores and RCI

	Random			*Agroecological fairs*		
	Ibarra	*Quito*	*Riobamba*	*Ibarra*	*Quito*	*Riobamba*
Salt control practices						
Mean (Std)	0.32 (0.62)^{aA}	0.68 (0.91)^{bA}	0.59 (1.23)^{aA}	0.46 (0.65)^{aA}	0.74 (0.98)^{aA}	0.87 (1.40)^{aB}
Median	0.00	0.00	0.00	0.00	0.00	0.00
(Min–Max)	(0–3)	(0–6)	(0–6)	(0–2)	(0–5)	(0–6)
Correlation between salt control practices and RCI						
Spearman correlation	0.07	0.15	0.07	0.13	0.14	0.19
p	0.008	<.0001	0.04	0.39	0.001	0.001
n	1284	769	861	48	551	299

*adjusted according to sampling weights.

food among their population. Here we explore some of the potential reasons for this heterogeneity across counties.

The higher RCI scores for households in Riobamba were driven by the higher scores for Dimensions One (direct purchase from producers) and Two (preference for agroecological produce). The higher value for Dimension Two in Riobamba is consistent with the higher proportion of open markets per capita in Riobamba, where there is 1 open market space per 5,641 families, compared to 1 per 8,111 in Ibarra, and 1 per 19,417 in Quito (Ekomer, 2017).

However, in the group sampled at agroecological fairs, Quito had the highest value for the three dimensions. This likely reflects the strong awareness-raising process of the '250 thousand families', a citizens' campaign (www.quericoes.org) which promotes practices of responsible food consumption, focussing on populations involved in agroecological production and direct markets. This might also explain why a higher percentage of consumers in fairs in Quito also consume fruits and vegetables.

Some of the differences observed between the random sample and the fairs sample are due to the nature of the variables considered for Dimensions One and Two. Indeed, as direct and agroecological purchasing sites are often the same, it is expected that consumers sampled in agroecological fairs would obtain a higher score for these dimensions and that they should be more highly correlated. That this is the case is partial validation of the dimensions and RCI.

For Dimension Three, on Andean grains, the Ibarra random sample had a higher score than the other counties. This finding is consistent with Ibarra being a centre of quinoa (Subsecretaría de Agricultura, 2015) and chocho (Peralta, 2016) production, likely influencing consumption among the general population. Nevertheless, agroecological fairs seem to be an effective way of influencing Andean grains consumption, as shown by the high score found in the Quito fairs sample.

Practices aimed at regularly and actively controlling table salt in the diet were uncommon in all counties and both samples, making this potential link between responsible consumption and a nutritionally relevant practice hard to make (with low rank correlations). The positive association between responsible food consumption and (some aspects of) the quality of the diet (that is, fruit and vegetable consumption, but not salt control) is intriguing. Does practising responsible consumption lead to greater fruit and vegetable consumption (perhaps by directing the shoppers to markets where fruits and vegetables are sold)? Or do health concerns increase the pursuit of fruit and vegetables and lead consumers to agroecological markets, which increases the RCI? Whatever the nature of the relationship, it may be expected that if food consumption in Ecuador becomes more 'responsible', it would also become consistent with public health promotion of fruit and vegetable consumption.

Conclusions and further research

The RCI represents a valuable tool for characterizing different city region populations and their food procurement and consumption patterns (see also Chapters 9 & 11, this volume). The RCI was also useful for establishing a concrete relationship between a more general consumption pattern and some specific behaviour in another domain, such as the consumption of fruits and vegetables. From a methodological perspective, despite the limitations of the RCI, its potential to compare patterns of food consumption with other environmental, economic, social, health, demographic, or other variables opens up a range of possibilities for the study of responsible consumption. The RCI represents both a relatively quick assessment tool and a starting point for further quantitative and qualitative research.

On the other hand, the nature of the quantitative data did not tap into consumer's intentions behind their responsible consumption practices. It may be that a consumer intends to consume responsibly, but does not demonstrate practices that contribute to a sustainable food system. Several barriers (e.g. physical, economic, temporal or cognitive) can explain this gap between intentions and behaviours, including the role of accepted sets of practices, termed by Lahlou (2018) as 'installations'. Conversely, a consumer who contributes to a sustainable food system through his practices may not necessarily be intentional, where the concept of responsible consumption implies a certain awareness, and active choices. Food consumption patterns are motivated by multiple factors and the intention to consume responsibly may not be the one that has motivated a practice that contributes to a sustainable food system. It may be because of the organoleptic qualities of food, the proximity of markets, the incentives of a given public policy, or other reasons. Studies evaluating consumer intentions or combining questions about intentions with observations on actual practices would help to inform the conceptualization and operationalization of responsible food consumption.

The RCI could be useful for testing how the food environment influences food consumption patterns. Shoppers at agroecological fairs tended to have higher responsible consumption indexes. In terms of public policy, this suggests that agroecological markets should be promoted and supported to give more consumers the opportunity to choose responsibly for their food system. Citizens can influence their food environment through campaigns and organizations that promote the creation of neighbourhood, open and agroecological markets, as was shown by the data on Quito. Further applications of the RCI in different contexts may elucidate other patterns and explore different opportunities to understand responsible consumption's contribution to sustainable food systems and better health.

Acknowledgements

The Ekomer team includes Malek Batel, Pablo Lopez, and Stephen Sherwood. We also thank survey participants and interviewees.

Note

1 BMI, Body Mass Index, is calculated as weight (in kg) divided by the square of height (in m). A BMI of 25 to 30 is considered overweight, and a BMI over 30 is considered obese. Available at: http://www.who.int/en/news-room/fact-sheets/detail/obesity-and-overweight. Accessed 22 October 2018.

References

Allen, L.N. & Feigl, A.B. (2017) Reframing non-communicable diseases as socially transmitted conditions. *The Lancet Global Health*. 5(7), pp. 644–646.

Altieri, M. (1999) Applying agroecology to enhance the productivity of peasant farming systems in Latin America. *Environment Development and Sustainability*. 1(3), pp. 197–217.

Anderson, W.T. & Cunningham, W.H. (1972) The socially conscious consumer. *The Journal of Marketing*. 36(3), pp. 23–31.

Antil, J. H. (1984) Socially responsible consumers: Profile and implications for public policy. *Journal of Macromarketing*. 4(2), pp. 18–39.

Beccaria, S. (2016) *Evènement parallèle: Présentation des actions de Slow Food en Méditerranée*. Report available at: http://www.fao.org/in-action/quality-and-origin-program/en and https://www.slideshare.net/ExternalEvents/evnement-parallle-prsentation-des-actions-de-slow-food-en-mditerrane-s-beccaria-english.

Berkowitz, L. & Lutterman, K.G. (1968) The traditional socially responsible personality. *Public Opinion Quarterly*. 32(2), pp. 169–185.

Blay-Palmer, A. (2006) Imagining sustainable food system. In: Blay-Palmer, A. (Ed.) *Imagining sustainable food system: Theory and practice*. London, Routledge.

Brug, J. (2008) Determinants of healthy eating: Motivation, abilities and environmental opportunities. *Family Practice*. 25(1), pp. 50–55.

Burlingame, B. & Dernini, S. (2012) Sustainable diets and biodiversity: Directions and solutions for policy, research and action. *Proceedings of the International Scientific Symposium Biodiversity and Sustainable Diets United Against Hunger*. 3–5 November, 2010. Rome, FAO.

Carrion, D. & Herrera, S. (2012) *Ecuador rural del siglo XXI. Soberanía alimentaria, inversión pública y política agraria*. Instituto de Estudios Ecuatorianos. Quito, Ecuador.

CEPAL [Comisión Económica para América Latina] (2014) *Agricultura familiar y circuitos cortos: Nuevos esquemas de producción, comercialización y nutrición*. CEPAL, Serie Seminarios y Conferencias (77).

Cerdán, C. (2014) La experiencia de los Sistemas Agroalimentarios Localizados (SIAL) en América Latina. In: *Agricultura familiar y circuitos cortos: Nuevos esquemas de producción, comercialización y nutrición*. CEPAL, Serie seminarios y conferencias.

Chauveau, C. & Taipe, D. (2012) *Circuitos alternativos de comercialización*. Quito, Ecuador, Ministerio de Agricultura, Ganadería Acuacultura y Pesca del Ecuador, AVSF.

Cole, D.C., Sherwood, S., Paredes, M., Sanin, L.H., Crissman, C., Espinosa, P., & Muños, F. (2007) Reducing pesticide neurotoxic effects in farm households. *Internatonal Journal of Occupational and Environmental Health*. 13(3), pp. 281–289.

Contreras, J., Paredes, M., & Turbay, S. (2018) Circuitos cortos de comercialización agroecológica en el Ecuador. *Idesia*. 35(3), pp. 71–80.

Craviotti, C. & Soleno Wilches, R. (2015) Circuitos cortos de comercialización agroalimentaria: Un acercamiento desde la agricultura familiar diversificada en Argentina. *Mundo Agrario*. 16(33).

Csutora, M. & Vetöné, M. (2014) Consumer income and its relation to sustainable food consumption – obstacle or opportunity? *International Journal of Sustainable Development & World Ecology*. 21(6), pp. 512–518. Available at: http://dx.doi.org/10.1080/13504509.2014.965238.

Dixon, J. (2009) From the imperial to the empty calorie: How nutrition relations underpin food regime transitions. *Agriculture and Human Values*. 26(4), pp. 321–333.

Dueñas Ocampo, S., Perdomo-Ortiz, J., & Villa Castaño, L.E. (2014) El concepto de consumo socialmente responsable y su medición. Una revisión de la literatura. *Estudios Gerenciales*. 30(132), pp. 287–300.

Ekomer (2017) Database of fairs and other agroecological shops in Quito, Ibarra, and Riobamba counties (unpublished).

ENIGHUR-INEC (2013) *Encuesta nacional de ingresos y gastos de hogares urbanos y rurales (ENIGHUR 2011–2012)*. Available at: http://www.ecuadorencifras.gob.ec/encuesta-nacional-de-ingresos-y-gastos-de-los-hogares-urbanos-y-rurales/.

ENSANUT-ECU (2012) *Encuesta nacional de salud y nutricion*. Available at: http://www.ecuadorencifras.gob.ec/documentos/web-inec/Estadisticas_Sociales/ENSANUT/MSP_ENSANUT-ECU_06-10–2014.pdf.

FAO (2010) *Sustainable diets and biodiversity directions and solutions for policy, research and action*. Rome, FAO. Available at: http://www.fao.org/docrep/016/i3004e/i3004e.pdf.

FAO (2018) *The 10 elements of agroecology. Guiding the transition to sustainable food and agricultural systems*. Rome, FAO. Available at: http://www.fao.org/3/I9037EN/i9037en.pdf.

Feldman, P.M. & Reficco, E. (2015) Impacto de la responsabilidad social empresarial en el comportamiento de compra y disposición a pagar de consumidores bogotanos. *Estudios Gerenciales*. 31(137), pp. 373–382.

Francois-Lecompte, A. (2005) *La consommation socialement responsable: Proposition et validation d'un cadre conceptuel intégrateur*. These pour l'Obtention du Doctorat en Sciences de Gestion. Universite Pierre Mendes France, Ecole Doctorale de Sciences de Gestion. Grenoble, France. Available at: https://www.researchgate.net/profile/A_Francois-Lecompte/publication/281876972_La_consommation_socialement_responsable_proposition_d'un_cadre_d'analyse_integrateur/links/57e231d008ae9e25307d56c3/La-consommation-socialement-responsable-proposition-dun-cadre-danalyse-integrateur.pdf.

Fraňková, E. & Haas, W. (2017) *Socio-metabolic perspectives on the sustainability of local food systems*. Cham, Springer.

Freire, W.B., Ramírez, M.J., Belmont, P., Mendieta, M.J., Silva, M.K., Romero, N., Sáenz K., Piñeiros, P., Gómez, L.F., & Monge, R. (2013) *Resumen ejecutivo. Tomo I. Encuesta Nacional de Salud y Nutrición del Ecuador*. Quito, Ecuador, ENSANUT-ECU 2011–2013 Ministerio de Salud Pública/Instituto Nacional de Estadística y Censos.

Freire, W.B., Silva-Jaramillo, K.M., Ramírez-Luzuriaga, M.J., Belmont, P., & Waters, W.F. (2014) The double burden of undernutrition and excess body weight in Ecuador. *The American Journal of Clinical Nutrition*. 100(6), pp. 1636S–1643S.

Frynas, J.G. & Yamahaki, C. (2016). Corporate social responsibility: Review and roadmap of theoretical perspectives. *Business Ethics: A European Review*. 25(3), pp. 258–285.

Fundación Heifer (2014) *La agroecología está presente. Mapeo de productores agroecológicos y del estado de la agroecología en la sierra y costa ecuatoriana.* Quito, Ecuador, Fundación Heifer Ecuador y Ministerio de Agricultura, Ganadería, Acuacultura y Pesca (MAGAP).

GBD [Global Burden of Disease Obesity Collaborators] (2017) Health effects of overweight and obesity in 195 countries over 25 years. *New England Journal of Medicine*. 377(1), pp. 13–27.

Gliessman, S.R. (2007) *Agroecology: The ecology of sustainable food systems*. 2nd Edition. Boca Raton, FL, CRC Press, Taylor & Francis Group.

González, I., De Haro, T., Ramos, E., & Renting, H. (2012) Circuitos cortos de comercialización en Andalucía: Un análisis exploratorio. *Revista Española de Estudios Agrosociales y Pesqueros*. 232, pp. 193–227.

Gould, S.J. (1988) Consumer attitudes toward health and health care: A differential perspective. *Journal of Consumer Affairs*. 22(1), pp. 96–118.

Government of Ecuador (2010) *Ley Orgánica del Régimen de la Soberanía Alimentaria*. Available at: https://www.soberaniaalimentaria.gob.ec/pacha/wp-content/uploads/2011/04/LORSA.pdf.

Guerrón-Montero, C. & Moreno-Black, G. (2001) Household structure and dietary patterns in the Afro-Ecuadorian highlands. *Food and Nutrition Bulletin*. 22(1), pp. 23–30.

Gustafson, D., Gutman, A., Leet, W., Drewnowski, A., Fanzo, J., & Ingram, J. (2016) Seven food system metrics of sustainable nutrition security. *Sustainability*. 8, pp. 196–213.

Guthman, J., Broad, G., Klein, K., & Landecker, H. (2014) Beyond the sovereign body. *Gastronomica: The Journal of Food and Culture*. 14(3), pp. 46–55.

Heinisch, C. (2016) Conceptualizing and creating sustainable food systems: How interdisciplinarity can help. In: Blay-Palmer, A. (Ed.) *Imagining Sustainable Food Systems: Theory and Practice*. London, Routledge.

HLPE [High Level Panel of Experts on Food Security and Nutrition] (2014) *Food losses and waste in the context of sustainable food systems*. A report by the HLPE of the Committee on World Food Security, Rome. Available at: www.fao.org/cfs/cfs-hlpe.

INEC [Instituto Nacional de Estadisticas y Censo] (2010) *Censo de población y vivienda*. Available at: http://www.ecuadorencifras.gob.ec/censo-de-poblacion-y-vivienda/.

INEC–ESPAC (2017) *Encuesta de superficie y producción agropecuaria continua* (ESPAC). Available at: http://www.ecuadorencifras.gob.ec/documentos/web-inec/Estadisticas_agropecuarias/espac/espac_2017/Informe_Ejecutivo_ESPAC_2017.pdf. Accessed 12 August 2018.

Inglehart, R. (2015) *The silent revolution: Changing values and political styles among western publics*. Princeton, NJ, Princeton University Press.

Jacobsen, S. & Sherwood, S. (2002) *Cultivo de granos andinos en Ecuador. Informe sobre los rubros quinua, chocho y amaranto*. Organicación de las Naciones Unidas para la Agricultura (FAO), Centro Internacional de la Papa (CIP), y Catholic Relief Services (CRS). Quito, Ecuador.

Kaplan, R.M., Bush, J.W., & Berry, C.C. (1976) Health status: Types of validity and the index of well-being. *Health Services Research*. 11(4), pp. 478–507.

Kinnear, T.C., Taylor, J.R., & Ahmed, S.A. (1974) Ecologically concerned consumers: Who are they? *Journal of Marketing*. 38(2), pp. 20–24.

Lahlou, S. (2018) *Installation theory: The societal construction and regulation of behaviour.* Cambridge, Cambridge University Press.

Lang, T. & Heasman, M. (2015) *Food wars: The global battle for mouths, minds and markets*. Second edition. New York, Routledge.

Linders, H. (2014) Consumo ético, consumo saludable: Existe un nuevo consumidor en América Latina? In: *Agricultura familiar y circuitos cortos: Nuevos esquemas de producción, comercialización y nutrición*. CEPAL, Serie seminarios y conferencias.

López, S., Marín L., & Ruiz de Maya, S. (2017) Introducing personal social responsibility as a key element to upgrade CSR. *Spanish Journal of Marketing- ESIC*. 21(2), pp. 146–163.

Maas, V. (2017) Observing responsible food consumption in families living in Ecuador. Getting closer to food sovereignty with civil action? Unpublished Master's thesis. Humboldt Universität Zu Berlin.

MAGAP [Ministerio de Agricultura, Ganadería, Acuacultura y Pesca] (2016) *La política agropecuaria Ecuatoriana: Hacia el desarrollo territorial sostenible: 2015–2025 I Parte*. Quito, Ecuador.

Mertens, E., Van't Veer, P., Hiddink, G.J., Steijns, J.M., & Kuijsten, A. (2016) Operationalising the health aspects of sustainable diets: A review. *Public Health Nutrition*, 20(4), pp. 1–19.

MIES, CEPAL & Programa Mundial de Alimentos (2017) *Impacto social y económico de la malnutrición. Resultados del estudio realizado en Ecuador*. Available at: http://es.wfp.org/sites/default/files/es/file/el_costo_de_la_doble_carga_de_la_malnutricion_-_ecuador.pdf.

Ministerio de Cultura y Patrimonio (2013) *Patrimonio alimentario*. Quito, Ecuador, Fasículo 1.

Moorman, C. & Matulich, E. (1993) A model of consumers' preventive health behaviors: The role of health motivation and health ability. *Journal of Consumer Research*. 20(2), pp. 208–228.

Muzo, A., Belmont, P., & Navarrete, I. (2017) *Análisis de los circuitos cortos de los productos agroecológicos en el Ecuador*. Ecuador, AllpaData, pp. 1–55.

Newholm, T. & Shaw, D. (2007) Studying the ethical consumer: A review of research. *Journal of Consumer Behaviour*. 6(5), pp. 253–270.

Ottman, J.A. & Reilly, W.R. (1998) *Green marketing: Opportunity for innovation*. 2nd edition. Upper Saddle River, NJ, Prentice Hall.

Ozçaglar-Toulouse, N. (2009) What meaning do responsible consumers give to their consumption? An approach by narratives. *Recherche et Applications en Marketing* (English Edition). 24(3), pp. 3–20.

Paredes, M. (2010) Peasants, potatoes and pesticides. Heterogeneity in the context of agricultural modernization in the highland Andes of Ecuador. Unpublished Ph.D. Thesis. Wageningen University and Research Center, The Netherlands.

Peralta, E. (2016) *El chocho en Ecuador. Estado del Arte*. Available at: http://repositorio.iniap.gob.ec/bitstream/41000/3938/1/iniapscdpCD99.pdf.

Peralta, E., Mazón, N., Murillo, A., Rivera, M., Rodríguez, D., Lomas, L., & Monar, C. (2012) *Manual agrícola de granos Andinos: Chocho, quinua, amaranto y ataco.*

Cultivos, variedades y costos de producción. 3rd ed., No. 69. Programa Nacional de Leguminosas y Granos Andinos. Estación Experimental Santa Catalina. INIAP. Quito, Ecuador.

Piñeiro, C. & Díaz, M.J. (2012) Lost in the city: Responsible consumption as message, and the city of Madrid as communication ecosystem. *Athenea Digital*. 12(1), pp. 67–88.

Pivato, S., Misani, N., & Tencati, A. (2008) The impact of corporate social responsibility on consumer trust: The case of organic food. *Business Ethics: A European Review*. 17(1), pp. 3–12.

Roberts, P., KC, S., & Rastogi, C. (2006) *Rural access index: A key development indicator*. Washington, DC, The World Bank Group.

Sarandón, S.J. & Flores, C.C. (Eds.) (2014) *Agroecología: Bases teóricas para el diseño y manejo de agroecosistemas sustentables*, 1st ed. La Plata, Universidad Nacional de La Plata.

Sherwood, S. (2009) Learning from Carchi. Agricultural modernization and the production of decline. Unpublished Ph.D. Thesis. Wageningen University and Research Center, The Netherlands.

Sonnino, R. (2009). Feeding the city: Towards a new research and planning agenda. *International Planning Studies*. 14(4), pp. 425–435.

Stern, P.C. & Dietz, T. (1994) The value basis of environmental concern. *Journal of Social Issues*. 50(3), pp. 65–84.

Subsecretaría de Agricultura (2015) *Datos Referenciales*. Quito, Ecuador, Ministerio de Agricultura, Ganadería, Acuacultura y Pesca.

Suquilanda, M. (2006) *Agricultura orgánica. Alternativa tecnológica del futuro*. 3rd ed. Quito, Ecuador, Fundación para el desarrollo agropecuario.

Tsai, Y.H., Joe, S.W., Lin, C.P., Chiu, C.K., & Shen, K.T. (2015) Exploring corporate citizenship and purchase intention: Mediating effects of brand trust and corporate identification. *Business Ethics: A European Review*. 24(4), pp. 361–377.

Unigarro Solarte, C. (2010) *Patrimonio cultural alimentario*. Quito, Ecuador, Fondo Editorial Ministerio de Cultura. Cartografía de la memoria no 4.

Villacrés, E., Navarrete, M., Lucero, O., Espín, S., Peralta, E. (2010) Evaluación del rendimiento, características físico-químicas y nutraceúticas del aceite de chocho (Lupinus mutabilis sweet). *Revista tecnológica ESPOL-RTE*. 23(2), pp. 57–62.

Warde, A. (2005) Consumption and theories of practice. *Journal of Consumer Culture*. 5(2), pp. 131–153.

Webster, F.E. (1975) Determining the characteristics of the socially conscious consumer. *Journal of Consumer Research*. 2(3), pp. 188–196.

Yépez, R., Carrasco, F., & Baldeón, M. (2008) Prevalencia de sobrepeso y obesidad en estudiantes adolescentes ecuatorianos del área urbana. *Archivos Latinoamericanos de Nutrición*. 58(2), pp. 139–143.

Zimmer, M., Stafford, T., & Stafford, M. (1994) Green issues: Dimensions of environmental concern. *Journal of Business Research*. 30(1), pp. 63–74.

11 Integrating upstream determinants and downstream food metrics

Nevin Cohen

Introduction

Planners acknowledge that urban food systems should be measured and managed as complex, adaptive systems, interconnected sets of dynamic social, physical, economic, and cultural phenomena (Meter, 2010; see also Chapters 1, 4, & 12, this volume). A recent review identified 260 distinct food system indicators included in the food strategies and plans of five North American cities: New York, Philadelphia, Los Angeles, Chicago, and Toronto (Coppo et al., 2017; Ilieva, 2017). NGOs and governments have also developed multidimensional indicators to track food system governance, diet-related public health outcomes, and the environmental impacts of urban food systems (Prosperi et al., 2015; Milan Urban Food Policy Pact, 2017; see also Chapter 7, this volume). Despite efforts to incorporate food metrics that measure the root causes of downstream food system outcomes in food plans, such as poverty and discrimination, development pressures, or labour exploitation, these remain exceptions, not the norm. More typically, cities collect a narrower range of metrics to manage their food systems that are oriented to downstream outcomes. Variables like food infrastructure (e.g., the distribution of food retailers), programme outputs (e.g., participation in public feeding programmes), population dietary practices (e.g., fruit and vegetable consumption and obesity prevalence), agency adherence to food policies (e.g., compliance with nutritional standards) are commonly tracked.

The disconnect between upstream determinants of the food system and downstream policies and programmes is certainly not unique to food policy. In the field of public health, for example, practitioners typically design interventions to change individual behaviours rather than addressing the upstream social determinants of those behaviours, like poverty, housing affordability, education, or environmental conditions (Freudenberg et al., 2015). The obstacles to integrating the upstream and downstream in food planning are conceptual, pragmatic, and political. The causal links between upstream determinants and downstream outcomes of the food system are long and complex, involving multiple intervening and interacting factors, making indicator development difficult (Braveman et al., 2011). Emphasizing the complexities of the food system can make it seem more difficult to frame solutions, and thus harder to rally support behind new initiatives, though Meter (Chapter 4, this volume) argues that taking complexity into account can actually facilitate consensus. The lack of adequate resources is a pragmatic constraint that prevents cities from tracking data on the social and economic variables that affect the food system, especially if there is only a modest budget for food systems planning and no mandate for city agencies to track these metrics. Other pragmatic factors include: professional traditions in fields like health and planning that focus staff on downstream interventions; the demand for short-term, measurable changes by administrations that favour quick results; silos among government officials, advocates, researchers, and funders that make data sharing difficult; and what Libman (2015) describes as a 'local trap' that emphasizes interventions within smaller rather than larger geographies. City officials may also be reluctant to draw attention to the politically fraught, 'wicked' problems (Rittel & Webber, 1973) of racial, class, or gender oppression, preferring to measure more discrete intervention outcomes, preferably those that demonstrate success, not failure. Finally, though food can be an effective way to teach about and address systemic inequities (Reynolds & Cohen, 2016), NGOs grappling with issues like immigrant rights, housing access, or fair labour practices may not recognize the potential benefits of framing their political strategies in the context of food.

Failing to integrate upstream determinants of food systems in the indicators used to manage urban food systems can affect food planning and policy development in several ways. As socially constructed representations of reality, metrics drive decision-making by their ability to frame problems, privilege certain analytical methods, and thus exclude potential solutions (Barrett, 2010; Sébastien & Bauler, 2013). They shape policies by reinforcing or contradicting established positions, suggesting positive or negative trends, and motivating or discouraging stakeholder involvement in policymaking (Hezri & Dovers, 2006). The process of identifying appropriate metrics and analysing their meanings can therefore facilitate shared understandings of problems and desired changes by engaging different actors, or present a limited or distorted view of reality, thus excluding stakeholders by constraining measurement to specific outcomes (Innes, 1990). Omitting upstream indicators like poverty or discrimination from food planning also elevates the importance of measured downstream interventions, like supermarket density, potentially diverting attention from, and quelling demands for, more radical social and political reforms like raising the minimum wage or capping commercial rents (Rosenberg & Cohen, 2017). Tracking only the downstream outputs of food programmes, like the number and productivity of urban farms, can mask the roles such programmes play in addressing upstream issues like governance within the food system, and by doing so minimize their transformative potential and dampen political and financial support for these projects (Reynolds & Cohen, 2016). Documenting food policy outcomes that address upstream determinants, like poverty reduction or increased social cohesion, can strengthen the case for such policies.

This chapter discusses the benefits of integrating upstream and downstream food metrics by illustrating how integration can focus policy-making on the root causes of three politically salient food policy problems in New York City: food insecurity among immigrants, unhealthy neighbourhood food environments, and poor labour conditions faced by food workers. Following a brief review of food metrics development in New York City, the chapter discusses how measuring upstream determinants of each issue can facilitate the design of more effective food policies and better equip food advocates with an understanding of the structural problems they need to solve. The chapter concludes with strategies that planners can use to more effectively and efficiently collect upstream metrics and integrate them into food planning (see Chapter 12, this volume).

The emergence of NYC food metrics

Cities in the Global North have collected data on food production and distribution since their founding, but surveillance of the food system, from adulteration and safety to food distribution infrastructure mapping, increased with the emergence of municipal planning and public health departments at the turn of the last century (Vitiello & Brinkley, 2014). But until the past few decades, and the proliferation of distinct food system plans, municipal agencies had not developed and collected urban food metrics systematically (Coppo et al., 2017; Ilieva, 2017). In New York City, for example, food metrics had not been aggregated and presented cohesively until the start of annual food metrics reporting in 2011 (Freudenberg et al., 2018). City agencies in charge of Health, Sanitation, Parks, Economic Development, and other agencies had published data for many years about the food programmes under their jurisdictions, such as the quantity of discarded organic material disposed of by the Sanitation Department, yet those metrics had never been compiled as food system indicators until food gained recognition as an urban system in need of planning, measuring, and managing.

The New York City Council's 2010 release of FoodWorks, a comprehensive food systems strategy, provided the impetus for tracking food metrics, as it was followed by the enactment of three local laws to require the city to collect and report food system data (Cohen, 2011). The metrics mandated by the City Council reflected food planning objectives of different advocates. One bill required the agency in charge of city property to publish a list of all city-owned vacant parcels with an assessment of their suitability to grow food, responding to urban agriculture proponents who wanted to expand food production.[1] Advocates for using the city's food purchasing power to support regional agriculture enabled enactment of a second bill requiring an annual report of New York State-produced food procured by city agencies.[2]

The third food metrics bill was designed to measure progress towards multiple objectives in the City Council's FoodWorks plan, requiring reporting of 37 indicators (subsequently amended to add food insecurity metrics) of food-related activities under the jurisdiction of different city agencies (New York City Council, 2013).[3] In determining the scope of the food metrics legislation, the Administration and City Council negotiated which data was deemed useful for food planning and feasible for existing staff to collect without significant additional resources (Campbell, 2016). The Office of Food Policy compiles metrics from different agencies and publishes them in an annual report.

The food metrics mostly (21 of 37 metrics) measure nutritional health, which was the focus of the Bloomberg administration and an area fully within the purview of the Department of Health and Mental Hygiene (DOHMH). The remaining metrics report on food insecurity (n=4), food-related economic development (n=3), food system environmental impacts (n=8), and the number of food workers trained by the city (n=1) (Freudenberg et al., 2018). An analysis of the values of each metric between the first food metrics report issued in 2012 and the report issued in 2017 showed varied changes in indicator outcomes. Between 2012 and 2017, 19 indicators showed improvements, 15 showed declines, one didn't change, and two were not able to be assessed. For example, the percentage of New York City residents reporting food insecurity fell by 14 per cent during this period, while the number of permits for Green Cart vendors who sell produce from mobile carts in low-income neighbourhoods declined by 37 per cent (Freudenberg et al., 2018; Freudenberg et al., 2018a).

In addition to the metrics published in the city's annual food metrics reports, many other New York City agencies collect and report data on the food system that are relevant to their missions, but these are disseminated through multiple agency websites and documents. For example, the DOHMH tracks prevalence of non-communicable diet-related diseases and publishes the results of restaurant and school cafeteria inspections. The Human Resources Administration estimates the percentage of those eligible who participate in the federal Supplemental Nutrition Assistance Program (SNAP), which provides supplemental money to buy food. The Department of Sanitation collects and reports on food waste management. Still other agencies that address land use (e.g., Department of City Planning), economic development (e.g., Economic Development Corporation), and neighbourhood redevelopment (e.g., Housing Preservation and Development) monitor and prepare reports on food distribution, manufacturing, and retail.

Moving upstream

The data presented in the city's annual food metrics report, along with the additional food system metrics compiled by independent agencies, address downstream outcomes of the food system, like dietary changes or programme results such as the value of New York food products procured by city agencies. Integrating upstream and downstream factors can provide a more complete picture of the factors influencing food policies and their outcomes, as illustrated by three examples from New York City: the effects of federal immigration policies on participation in SNAP within immigrant communities; real estate development pressures and neighbourhood food environments; and labour policies that affect the well-being of food workers.

Food insecurity among immigrant communities

To track progress in addressing food insecurity, New York City reports the number of people who indicate that in the previous year they faced insufficient access to food for an active, healthy life. In 2016, 1.22 million New Yorkers, 14.4 per cent of the population, reported being food insecure. The city also reports a related metric, an estimated annual 'meal gap' that indicates the number of meals foregone by households because of insufficient income. In 2016, New York City residents missed 207.7 million meals due to their inability to afford adequate food (New York City Mayor's Office of Food Policy, 2018). Reducing the percentage of city residents reporting food insecurity and the size of the meal gap are important policy goals. One strategy is ensuring that eligible New Yorkers participate in SNAP. The number of people participating in SNAP is an indicator of both the need for food assistance due to poverty and the help that New Yorkers receive through the programme to feed their households. The food metrics report presents the number of SNAP participants for the city, by Borough, and among older adults (New York City Mayor's Office of Food Policy, 2018).

Between 2012 and 2016, the annual SNAP participation rate declined by 7.3 per cent, although nearly 20 per cent of the city's population still receive SNAP benefits (Hunger Free America & CUNY Urban Food Policy Institute, 2018). The overall decline is primarily correlated with the city's economic recovery from the Great Recession of 2007–2008 and a reduction in poverty and unemployment, but also reflects decisions by those eligible for SNAP not to apply or to leave the programme. In addition to understanding the upstream factors that cause a large percentage of New York City's population to be poor enough to qualify for SNAP, understanding the factors that inhibit or encourage people to apply for SNAP is key to designing interventions that increase participation and thus improve food security among those in need of supplemental financial support.

A study of SNAP caseload data during this period of decreasing participation found significant variation in total and per capita SNAP participation from neighbourhood to neighbourhood, with participation in some communities dropping by 28 per cent and others growing by 8 per cent (Hunger Free America & CUNY Urban Food Policy Institute, 2018). These differences indicated the need for closer scrutiny of the upstream determinants that may influence participation at the neighbourhood scale. Such factors may include the effects of forces like gentrification on the number of eligible residents in a neighbourhood, or demographic changes like an increased population of immigrants that can influence whether residents of a community choose to participate in SNAP even if their low income enables them to qualify.

Currently an estimated 220,000 New York City residents who are not citizens but who are here legally are receiving SNAP benefits, cash assistance, or both, yet a survey of 50 frontline staff of community-based organizations that serve food-insecure people found significant deterrents to SNAP participation among some immigrant communities. One potential obstacle was the language barriers faced by some of the approximately 17 per cent of New Yorkers who speak languages other than the nine the city translates for informational material and application forms (Vignola et al., 2018). A more complex upstream barrier to SNAP participation is fear of deportation due to recent federal immigration policies. Survey respondents reported that current and proposed federal policies, along with political rhetoric from elected officials, have stoked fears of deportation and have had a chilling effect, not only on applications for federal programmes like SNAP, but also on inquiries about food benefits among immigrants with whom they have interacted (Vignola et al., 2018). Some immigrants seeking US residency believe, incorrectly for now, that participation in SNAP will designate them a 'public charge', which could be used as a basis for deportation. A recently proposed rule that would broaden the definition of a public charge to include those receiving SNAP and other benefits has only intensified fears of participating in the programme, even among eligible immigrants (Health Affairs Blog, 2018).

Data on upstream determinants of participation in food benefit programmes such as language barriers (e.g., the percentage of those eligible for SNAP who speak languages other than those on government forms) and deportation fears (e.g., the percentage of eligible immigrants who fear that SNAP participation will put them at risk) would help cities isolate the causes of changing rates of participation within specific neighbourhoods and suggest methods to increase participation. While quantitative data on the prevalence and effects of the factors identified above may not affect anti-immigrant political positions, metrics tracking the experiences of immigrants applying for and participating in SNAP, including qualitative data on their perceptions of and responses to federal policies, could enhance arguments for political change and also facilitate the design of more effective interventions targeting specific immigrant populations. Metrics would also enable planners to anticipate the effects of future shifts in immigration policy on SNAP participation and develop alternatives to prevent hunger and malnourishment in immigrant communities. Surveying recent immigrants in the current anti-immigrant climate is a significant challenge for public agencies, even with supportive local governments, but it may be easier if agencies partner with non-profits who have the trust of immigrant communities.

Table 11.1 Marginal effects of federal food benefits on NYC poverty rate, in % change, 2012–2016

Program	*2012*	*2013*	*2014*	*2015*	*2016*
SNAP	-3.7	-4.0	-3.6	-3.2	-3.3
School Meals	-0.5	-0.6	-0.5	-0.7	-0.6
WIC	-0.3	-0.2	-0.2	-0.2	-0.3

Source: NYC Mayor's Office of Operations, 2018.

In addition to measuring the effects of upstream factors on SNAP participation and resulting downstream effects like reduced malnourishment and diet-related diseases, food planners can also track the upstream impacts to strengthen the case for the programme. For example, the New York City Mayor's Office of Operations analyses the effects of SNAP and two other federal food programmes (school meals and the Women, Infants, and Children programme), on the New York City poverty rate (NYC Mayor's Office of Operations, 2018). The most recent study shows that SNAP benefits reduce the poverty rate by more than 3 per cent, as Table 11.1 illustrates (NYC Mayor's Office of Operations, 2018). Tracking participation in SNAP not only as a way to address malnourishment but also as a tool for poverty alleviation provides evidence of the broader value of SNAP and illustrates the value of collaboration between advocates for health and food justice and anti-poverty advocacy groups.

Development and local food environments

A neighbourhood's spatial configuration, such as the availability of transit, neighbourhood services, and civic spaces for social interaction have long been viewed as important upstream determinants of health (Braveman et al., 2011). In recent years, researchers have focused on the local food environment, defined as the prevalence and configuration of food establishments, as factors in population nutrition and health (Malambo et al., 2016). Typically, food environments are measured based on the mix and location of food retailers, with the availability of convenience stores, corner stores, bodegas, and fast food restaurants proxies for unhealthy food environments and supermarkets as indicators of healthy food access (Lytle & Sokol, 2017; Rosenberg & Cohen, 2018). Researchers have also measured access to food in adjacent neighbourhoods and travel patterns to understand how food environments shape decisions about shopping, diets, and malnourishment (Lytle & Sokol, 2017). Upstream determinants of these patterns, such as development pressures and zoning decisions, are not typically measured as food metrics.

New York City's Food Metrics Reports track results of several policies to increase physical access to healthier food: the number of retailers participating in a programme called *Shop Healthy*, which helps bodegas and independent grocers sell healthier food; the number of Green Carts (mobile produce vendors selling in neighbourhoods underserved by supermarkets); and the number of new and expanded supermarkets supported by the Food Retail Expansion to Support Health (FRESH) policy, which offers incentives to increase grocery square footage in underserved neighbourhoods (New York City Mayor's Office of Food Policy, 2018). The metrics show that 36 FRESH supermarkets have received incentives to open or expand, but since 2011, 273 additional supermarkets opened without FRESH subsidies (New York City Mayor's Office of Food Policy, 2018). The number of Green Cart permits has declined to 286 (New York City Mayor's Office of Food Policy, 2018), though there has been no analysis of why there are so few Green Carts when 1,000 permits are available.

Incorporating upstream factors that determine the mix of food retailers in a community can more thoroughly explain changes in local food environments. Many variables determine retail development, from the health of the economy to the structure of retail sectors. At the community scale, zoning is an important determinant of real estate development potential. Zoning can restrict or induce commercial and residential development by changing the allowable uses, size, density, or configuration of buildings. Zoning changes determine food retailer locations and can spur displacement of existing food businesses by making other land uses more profitable (Cohen, 2018). Rezoning can also have secondary effects on food environments by stimulating overall real estate development, increasing population density, and attracting new, more affluent residents to a neighbourhood, potentially reducing the ratio of food retail per capita and encouraging retailers to market to the new higher income residents (Cohen, 2018). When zoning attracts wealthier residents to a low-income neighbourhood it can also lead to gentrification and displacement of existing residents.

Despite these direct and indirect effects of zoning on local food environments, cities rarely treat zoning changes as food system interventions and seldom analyse their potential consequences when they conduct public reviews. Because these effects on food are rarely documented in land-use review processes, they are infrequently raised by local advocates in public hearings. In New York City's environmental review process, for example, secondary or induced displacement of food retailers as a result of new commercial activity is not typically analysed, based on the assumption that commercial food establishments will open to meet any increased market demand. If environmental impact statements do not measure the adverse impacts of proposals on food access, they will not be identified and public reviews will likely overlook them.

An analysis of the effects of rezoning on the food environment of East Harlem, a historically low-income community of colour in upper Manhattan facing gentrification pressure, shows that upstream determinants of food environments like zoning-induced development pressures have changed the types of food retailers in the community and thus have affected access to healthy, affordable food. East Harlem has been rezoned numerous times since the late 1990s to stimulate economic development and in response to pressure from the real estate industry. The rezoning and subsequent development has had three principal effects on East Harlem's food environment: supermarket displacement; the creation of new sites for big box food retailers; and the expansion of both healthy and unhealthy food retail as the neighbourhood has gentrified (Cohen, 2018).

Rezoning Harlem's historic 125th Street made higher-density mixed-use development feasible and increased real estate values along this major commercial corridor. The increased land value enabled a local community development corporation to sell property it had previously acquired from the city at a below market price, with the purpose of developing an affordable supermarket called Pathmark (Cohen & Freudenberg, 2016). The company that purchased the Pathmark site now plans to build two 32-story residential towers on the land, yet has not committed to replacing Pathmark with another supermarket, despite demands from local elected officials and community leaders (Cohen & Freudenberg, 2016). A second example is the rezoning of an abandoned industrial site in East Harlem into a regional shopping centre designed to house big box food retailers that currently include Costco, Target, and Aldi. The environmental assessment's analysis of alternatives considered but rejected a conventional neighbourhood-serving supermarket and smaller commercial spaces as unfeasible (Cohen, 2018). A third example is the extensive rezoning of East Harlem to increase the scale of development sites to boost land value and development potential. Such rezoning throughout East Harlem has attracted residential development occupied by higher-income residents, increasing the value of residential and commercial real estate (Cohen, 2018). The influx of higher-income people has led to residential and commercial gentrification, including displacement of low-cost food retailers (Busà, 2014).

Considering the effects of zoning and other aspects of land development on the spatial configuration of food retailers would help planners anticipate and address the potential for new development to spur food gentrification, the process by which higher-income residents contribute to the displacement of affordable food establishments by higher-priced grocers and restaurants, or induce changes in the products offered by existing retailers, effectively displacing existing residents from the neighbourhood food environment by making them feel that the remaining food retail establishments are not offering the foods they desire and can afford (Pearsall & Anguelovski, 2016; Cohen, 2018). Treating data on real estate development trends (e.g., changes in commercial rents and commercial vacancies) and zoning proposals (e.g., square feet of additional commercial space permitted in a community) as food metrics can enable planners to anticipate the effects of these upstream factors on food retail and suggest strategies to prevent food gentrification. Strategies might include city financial support for existing food businesses, policies to prevent property owners from warehousing vacant commercial spaces, or the use of zoning incentives to attract affordable grocers and restaurants. Discussions about the determinants of food retail could be extended to include other upstream factors such as poverty or housing affordability, which affect the kinds of commercial activities that a community can support. Modifying the public review procedures that produce data-rich analyses of proposed developments, like environmental assessments and land-use review documents, so that they estimate the direct and secondary displacement of food retailers, would provide advocates with data on potential impacts and thus empower them to support land-use changes that enhance food access.

Good jobs for food workers

The workplace and working conditions are important upstream determinants of health. However, much of the occupational health and safety field has focused on the physical effects of work and environmental hazards in the workplace, rather than on the nature of employment, from the wages paid to workplace rules, that significantly affect a worker's economic status and emotional and physical health. In addition to salaries, workplace benefits (e.g., health insurance and paid leave) determine whether workers can afford healthier living conditions, including healthy, adequate food, yet those in low-wage jobs often earn too little to cover basic household needs. Low-wage jobs also prevent workers from having much control over their work processes, leading to irregular work schedules, insecure employment, and limited decision-making capacity that can create stress and other psychosocial impacts associated with the increased likelihood of injury, morbidity, and mortality, including diet-related chronic diseases (Lowden et al., 2010).

The conditions of food workers are particularly important to measure. The food sector has grown significantly over the past decade (Freudenberg et al., 2016). However, most of the food jobs that have been created in the United States since the Great Recession of 2007–2008 have been non-union, insecure, hourly labour in food services and food retail. In New York City, for example, the food sector is one of the largest and fastest growing job sectors, with 63,000 grocery store workers and 320,000 food service and drinking establishment employees, yet these jobs pay among the lowest wages of any employment sector (Freudenberg et al., 2016).

Over the past decade, New York City has adopted policies to address two important upstream determinants of nutritional health. One set of policies has increased wages for workers, and by doing so has increased incomes for low-wage food workers. A second set of policies has improved working conditions for the lowest-wage workers, which in New York City are concentrated in the food sector.

Higher wages

In 2012, the city enacted the Fair Wages for New Yorkers Act (Local Law 37 of 2012), which increased the living wage that businesses getting financial benefits from New York City have to pay their workers. Two years later, Mayor de Blasio signed an executive order expanding the coverage of this law to additional categories of workers and indexed the amount of the required living wage to the Consumer Price Index, raising the amount that the jobs covered by this law pay to approximately $15 per hour by 2019. The executive order covers an estimated 18,000 additional workers, 4,100 of which are in minimum wage jobs in the retail and fast-food sectors.

Over the past few years there has been a growing movement to raise the minimum wage, particularly for fast-food workers, buoyed by the national *Fight for Fifteen* movement, which calls for fast-food employers to provide at least a $15 hourly wage. Support by the Mayor and Governor led the New York State Wage Board on 20 May 2015 to raise the fast-food minimum wage to $15. The mayor also approved an increase to a $15 minimum wage, by the end of 2018 for all city employees and non-profit human services contractors. Raising the city's minimum wage means higher incomes for approximately 25 per cent of minimum wage earners and their families, affecting approximately 1.46 million workers throughout New York State.

Improved working conditions

In 2014, the city enacted legislation to expand mandatory paid sick leave to smaller businesses and added categories of family members (e.g., sibling, grandchild, and grandparent) for whom sick leave can be taken. These expansions in the new law extended sick leave coverage to an estimated 350,000 additional workers. Sick leave is particularly beneficial for low-wage workers, many of whom lack savings and thus face extreme hardships if they lose pay from being sick. This is particularly true of food service workers, a low-wage sector in which fewer than half of all workers had sick leave benefits before the law took effect (Rankin, 2012). Guaranteeing paid sick leave not only ensures that food workers and other low-wage employees are able to attend to their health without losing wages, but it also enables workers not to report to their workplaces ill, reducing health risks to co-workers and customers, especially important for people who handle food.

Another issue unique to restaurant workers is tipped wage theft. In November 2015 the city enacted Local Law 104 of 2015, which created an Office of Labor Standards to address this and other labour laws. The Office is required to educate employers; create public education campaigns regarding worker rights; collect and analyse statistics on violations; research and promote programmes about worker protections, education and safety; and conduct investigations, serve subpoenas, and impose civil penalties on businesses that do not comply with NYC's labour standards.

In 2017, New York City enacted several local laws to improve the work life of shift workers, a category that includes many fast food workers, by regulating the way their work schedules are set. Fast-food employers must provide work schedules two weeks in advance, pay premiums for changes made to work schedules, and offer open shifts to existing fast food employees. The laws also ban schedules that require workers to both close the business at night and reopen it first thing in the morning, and require employers to provide their employees with 72 hours advance notice of their work schedules.

To prevent prospective employers from discriminating against those who have a criminal record, Local Law 63 of 2015 prohibits any employer from inquiring about a job applicant's criminal history until after the employer makes the applicant a conditional offer of employment. This law is particularly important to address discrimination in arrests and sentencing, which has resulted in a disproportionate level of incarceration among African-American men, limiting their economic opportunity (Martin et al., 2015).

Transitions within the food retail sector prompted New York City to adopt Local Law 11 of 2016, which is designed to protect workers when a supermarket is sold to another company. The law requires grocery store owners that purchase existing grocery stores to retain the previous owner's employees for a period of 90 days after the business is purchased. After the 90-day transition period, the new employer must evaluate these employees and consider continuing their employment.

Metrics documenting compliance with some of these labour policies tracked by the Department of Consumer Affairs illustrate the challenges faced by low-wage workers. Food or drink service employees report being paid below the minimum wage at nearly three times the rate of retail employees (17.5 per cent vs. 6.7 per cent) (New York City Department of Consumer Affairs, 2017). A large percentage of low-wage workers in New York City report that they have experienced workplace violations in the previous week, including: 54 per cent with at least one pay-related violation; 69 per cent who were asked to do 'off-the-clock' work; 77 per cent with violations of overtime pay; and 21 per cent with minimum wage violations (New York City Department of Consumer Affairs, 2017). Low-wage immigrant workers in New York City, a particularly vulnerable group, experience minimum wage violations at more than twice the rate of non-immigrant low-wage workers (25 per cent vs. 12 per cent) (New York City Department of Consumer Affairs, 2017).

Treating labour data, particularly metrics on labour violations among low-wage workers, as relevant to the large food labour force, and integrating these data with metrics on malnourishment and diet-related health outcomes, could help to design interventions in the workplace that would improve compliance with labour laws and improve the health of low-income workers. Integrating labour and food metrics would also help to re-focus economic development policies from those that attract undifferentiated food sector jobs to those that aim to create food jobs that pay living wages and engage workers in businesses that produce and distribute healthier food (Freudenberg et al., 2016). It would also encourage needed research on the diet-related health impacts of stressful, unstable working conditions, an emerging area of public health scholarship (Laraia et al., 2017).

Strategies for integrating upstream and downstream food metrics

As the previous examples illustrate, it is difficult to transform the food system without connecting upstream variables that have substantial effects on downstream food system outcomes like food security and diet-related health. The notion that social determinants, like income, affect diets and health has become part of the common discourse on food policy, as public testimony from New York City Human Resources Administration Commissioner Steven Banks (2014) illustrates: 'The long-term solutions are clear. When New Yorkers can earn a living wage and find affordable housing, they will have the ability to obtain the food they need to prevent hunger…'. Given the complexity of measuring upstream variables, however, a key question is how food policymakers, planners, and advocates, can integrate them into the more commonly measured metrics of food policy outcomes. The examples above suggest three strategies: (1) aggregating, organizing, and analysing seemingly disparate data collected by different agencies as food metrics; (2) using innovative methods to collect relevant upstream data efficiently and cost-effectively, including using big data and crowd-sourcing techniques; and (3) including social, economic, and spatial indicators in food planning processes. Each of these is addressed below.

Using diverse datasets

Cities create troves of metrics prepared by the public and private sectors. Within city government, municipal agencies track the progress of their own programmes, monitoring public health, economic development, environmental characteristics, and many other dimensions of city life. Agencies record performance data on their activities, from inspections to licensing, that often involve the food system. Many existing public metrics like those on poverty, discrimination, educational attainment, housing affordability, or access to parks and active transportation are important upstream determinants of food and nutrition outcomes. These data can be used to answer important food systems questions, like how policies making the city more or less affordable for low-wage workers affects food security and hunger, the degree to which workplace or environmental stressors related to unhealthy diets disproportionately affect particular racial or socio-economic groups, and whether processes like gentrification and displacement affect neighbourhood food access.

Some of these data are published on agency websites, some are made available through data portals that an increasing number of cities have created (e.g., Open Data New York City), while others are collected for internal agency use and are not as easily accessed. Creating an inventory of food system-related metrics already available across agencies and other branches of city government, as well as relevant data collected by state and federal agencies, is the first step towards identifying relevant metrics and demonstrating ways to use them in food planning. In addition to aggregating available data, cities can change the requirements of existing assessment processes to require food system data to be collected and included in those analyses. Environmental impact assessments are one opportunity to collect relevant data on the relationship between zoning and development proposals and food environments, but cities are also required to collect and report social, economic, and infrastructure data to various federal agencies, from Housing and Urban Development to the Department of Transportation, that could be aggregated and analysed to better understand downstream food system outcomes. Ensuring that an expanded collection of food-related metrics is used requires alliances between food system advocates and groups that work on upstream issues like poverty, racial discrimination, immigrants' rights, affordable housing, and other social justice concerns. Academic institutions can support this collaborative work by showing how seemingly disparate data can be organized, analysed, and used for advocacy.

Big food data

A second strategy is to use innovative methods to collect relevant upstream data efficiently and cost-effectively. This includes using big data to measure behaviours at the population scale. For example, anonymized GPS data from mobile phone apps can be analysed to better understand how populations make choices about shopping and dining, and how daily activity patterns vary by demographic group, helping planners target programmes, like supermarket incentives, more effectively (see Athey et al., 2018). Aggregated purchasing data can be used to better understand how various social and economic characteristics influence what people buy and eat. Geospatial data like Google Street View can be analysed to better understand the environmental factors that influence activities like shopping in communities and can track commercial and residential changes over time that affect local food environments (Bader et al., 2017).

Food planning

A third strategy is to adopt a more expansive notion of food system planning that addresses the root causes of food system inequities: siting food production and distribution infrastructure to reduce poverty as well as improve efficiency; focusing economic development plans on sectors of the food industry that offer pathways to higher-wage jobs; protecting food businesses that cater to low-income residents as neighbourhoods are rezoned; and changing planning processes to more effectively involve community stakeholders with knowledge of social determinants and hard-to-reach demographic groups, like recent immigrants. Integrating upstream determinants of health into the types of issues that food planners address requires interdisciplinarity, and successful planning processes can break down barriers among administrative agencies and advocacy groups and foster interdisciplinary approaches to problem-solving (Corburn, 2009).

Moving beyond the food metrics typically tracked to monitor progress in addressing the health, social, economic, and environmental impacts of the food system requires identifying the root causes of downstream outcomes and figuring out ways to aggregate, organize, and analyse this information so it is useful to various stakeholders and city government. This can seem daunting to food planners with neither the resources nor the power to aggregate, organize, and analyse such data. Fortunately, integrating upstream and downstream metrics can be carried out iteratively, by starting with existing relevant data, using lessons from other big data projects, and engaging in a food system planning process that brings multiple stakeholders together to track a broader range of food metrics, spanning issues from poverty to social wages (housing, healthcare, education) to economic and environmental trends.

Notes

1 City of New York. Local Law 48 of 2011.
2 City of New York. Local Law 50 of 2011.
3 City of New York. Local Law 52 of 2011.

References

Athey, S., Blei, D., Donnelly, R., Ruiz, F., & Schmidt, T. (2018) Estimating heterogeneous consumer preferences for restaurants and travel time using mobile location data. *arXiv.* preprint. doi:1801.07826.

Bader, M.D., Mooney, S.J., Bennett, B., & Rundle, A.G. (2017) The promise, practicalities, and perils of virtually auditing neighborhoods using Google Street View. *The ANNALS of the American Academy of Political and Social Science*. 669(1), pp. 18–40.

Banks, S. (2014) *Testimony to the New York City Council General Welfare Committee hearing on hunger in New York*. New York, New York City Human Resources Administration. Available at: https://www1.nyc.gov/assets/hra/downloads/pdf/news/testimonies/2014/nov_2014/HungerHearing_2014.pdf

Barrett, C.B. (2010) Measuring food insecurity. *Science*. 327(5967), pp. 825–828.

Braveman, P., Egerter, S., & Williams, D.R. (2011) The social determinants of health: Coming of age. *Annual Review of Public Health*. 32, pp. 381–398.

Busà, A. (2014) After the 125th street rezoning: The gentrification of Harlem's Main Street in the Bloomberg Years. *Urbanities*. 4(2), pp. 51–68.

Campbell, L.K. (2016) Getting farming on the agenda: Planning, policymaking, and governance practices of urban agriculture in New York City. *Urban Forestry & Urban Greening*. 19, pp. 295–305.

Cohen, N. (2011) *FoodWorks legislative package scheduled for vote*. Urban Food Policy blog. Available at: https://urbanfoodpolicy.com/2011/07/27/foodworks-legislative-package-scheduled-for-vote/.

Cohen, N. (2018) Unintentional food zoning: A case study of east Harlem, New York. In: Cabannes, Y. & Marocchino, C. (Eds.) *Integrating food into urban planning*. London, UCL Press.

Cohen, N. & Freudenberg, N. (2016) *Creating healthy food access in a changing food retail sector: Invitation to a dialogue*. New York, CUNY Urban Food Policy Institute.

Coppo, G., Stempfle, S., & Reho, M. (2017) Urban food strategies and plans: Considerations on the assessment construction. *City, Territory and Architecture*. 4(1), p. 8.

Corburn, J. (2009) *Toward the healthy city: people, places, and the politics of urban planning*. Cambridge, MA, MIT Press.

Freudenberg, N., Cohen, N., Poppendieck, J., & Willingham, C. (2018) *Food policy in New York City since 2008: Lessons for the next decade*. New York, CUNY Urban Food Policy Institute.

Freudenberg, N., Franzosa, E., Chisholm, J., & Libman, K. (2015) New approaches for moving upstream: How state and local health departments can transform practice to reduce health inequalities. *Health Education & Behavior*. 42(1 Suppl.), pp. 46S–56S.

Freudenberg, N., Silver, M., Hirsch, L., & Cohen, N. (2016) The good food jobs nexus: A strategy for promoting health, employment, and economic development. *Journal of Agriculture, Food Systems, and Community Development*. 6(2), pp. 283–301.

Freudenberg, N., Willingham, C., & Cohen, N. (2018a) The role of metrics in food policy: Lessons from a decade of experience in New York City. *Journal of Agriculture, Food Systems, and Community Development*. 8(2 Suppl.), pp. 191–209.

Health Affairs Blog (2018) The health impact of the proposed public charge rules. 27 September. doi:10.1377/hblog20180927.100295.

Hezri, A.A. & Dovers, S.R. (2006) Sustainability indicators, policy and governance: Issues for ecological economics. *Ecological Economics*. 60(1), pp. 86–99.

Hunger Free America & CUNY Urban Food Policy Institute (2018) *SNAP declines yet hunger persists: NYC caseloads 2012–2016 and the need to ease access to benefits*. Available at: https://www.hungerfreeamerica.org/sites/default/files/atoms/files/Declining%20SNAP%20Participation%20Report%20June%202018.pdf.

Ilieva, R.T. (2017) Urban food systems strategies: A promising tool for implementing the SDGs in practice. *Sustainability*. 9(10), p. 1707.

Innes, J. E. (1990) *Knowledge and public policy: The search for meaningful indicators*. Piscataway, NJ, Transaction Publishers.

Laraia, B.A., Leak, T.M., Tester, J.M., & Leung, C.W. (2017) Biobehavioral factors that shape nutrition in low-income populations: A narrative review. *American journal of preventive medicine*. 52(2S2), p. S118.

Libman, K. (2015) Has New York City fallen into the local trap? *Public Health*. 129(4), pp. 310–317.

Lowden, A., Moreno, C., Holmbäck, U., Lennernäs, M., & Tucker, P. (2010) Eating and shift work: Effects on habits, metabolism, and performance. *Scandinavian Journal of Work, Environment & Health*. 36(2), pp. 150–162.

Lytle, L.A. & Sokol, R.L. (2017) Measures of the food environment: A systematic review of the field, 2007–2015. *Health & Place*. 44, pp. 18–34.

Malambo, P., Kengne, A.P., De Villiers, A., Lambert, E.V., & Puoane, T. (2016) Built environment, selected risk factors and major cardiovascular disease outcomes: A systematic review. *PloS One*. 11(11), e0166846.

Martin, S.A., Harris, K., & Jack, B.W. (2015) The health of young African American men. *Jama*. 313(14), pp. 1415–1416.

Meter, K. (2010) Metrics from the field: Letting food systems emerge. *Journal of Agriculture, Food Systems, and Community Development*. 1(1), pp. 23–26.

Milan Urban Food Policy Pact (2017) *Building opportunities for local food policies: Urban monitoring framework*. Gathering and Mayor's Summit. Available at: http://www.milanurbanfoodpolicypact.org/wp-content/uploads/2017/12/FAO_MUFPP-monitoring-framework-Valencia.pdf.

New York City Council (2011) *Report of the governmental affairs division*. Committee on governmental operations. 27 July.

New York City Council (2013) *Local Law to amend the New York city charter and the administrative code of the city of New York, in relation to assessing food security*. Available at: http://legistar.council.nyc.gov/LegislationDetail.aspx?ID=1520864&GUID=953D129F-2808-48E0-9E9D-D3C577E7DCFF.

New York City Department of Consumer Affairs (2017) *The state of workers' rights in New York City*. Available at: https://www1.nyc.gov/assets/dca/downloads/pdf/workers/StateofWorkersRights-Report-2017.pdf.

New York City Mayor's Office of Food Policy (2018) *Food Metrics Report 2018*.

New York City Mayor's Office of Operations (2018) *New York City government Poverty Measure 2005–2016*.

Pearsall, H. & Anguelovski, I. (2016). Contesting and resisting environmental gentrification: Responses to new paradoxes and challenges for urban environmental justice. *Sociological Research Online*. 21(3), pp. 1–7.

Prosperi, P., Moragues, A., Sonnino, R., & Devereux, C. (2015) *Measuring progress towards sustainable food cities: Sustainability and food security indicators*. Report of the ESRC financed Project 'Enhancing the Impact of Sustainable Urban Food Strategies'. Available at: http://sustainablefoodcities.org/getstarted/developingindicators.

Rankin, N. (2012) *Still sick in the city: What the lack of paid leave means for working New Yorkers*. New York City, Community Service Society.

Reynolds, K. & Cohen, N. (2016) *Beyond the kale: Urban agriculture and social justice activism in NYC*. Athens, GA, The University of Georgia Press.

Rittel, H.W. & Webber, M.M. (1973) Dilemmas in a general theory of planning. *Policy Sciences*. 4(2), pp. 155–169.

Rosenberg, N.A. & Cohen, N. (2017) Let them eat kale: The misplaced narrative of food access. *Fordham Urban Law Journal*. 45, p. 1091.

Sébastien, L. & Bauler, T. (2013) Use and influence of composite indicators for sustainable development at the EU-level. *Ecological Indicators*. 35, pp. 3–12.

Vignola, E., Ruiz-Navarro, P., & Freudenberg, N. (2018) *Expanding immigrant access to food benefits in New York City: Defining roles for city and state government*. New York, CUNY Urban Food Policy Institute.

Vitiello, D. & Brinkley, C. (2014) The hidden history of food system planning. *Journal of Planning History*. 13(2), pp. 91–112.

12 The view from here

A critical consideration of sustainable food system assessments

Alison Blay-Palmer, Damien Conaré, Ken Meter, and Amanda Di Battista

Introduction

This final chapter pulls together the perspectives that framed this book by considering the academic literature in light of the three overarching themes from the Toronto workshop. As elaborated in Chapter 1, the three themes are:

1. Conceptual foundations;
2. Operationalizing metrics;
3. Outcomes and goals for assessment projects.

These dimensions and their interconnections are captured in Figure 12.1, which provides a summary framework for *Sustainable Food Systems Assessment: Global Approaches to Practice*. As is expanded in the conclusion of this chapter, several key points emerge. First, the imperative to give voice to and/or reinforce socio-political processes founded in social movements and the evolving relationship between policymakers, practitioners, civil society, and academics. We also discuss what defines 'assessment', including the importance of stories, trust, and the social (including the social economy) as we link and identify indicators as nested, linked, *and* relational. The opportunity to render the invisible visible is a part of how assessment can bring challenges to the surface and transform solutions through understanding, transparency, and building on food system complexity.

The research reported in this book points to assessments as not only providing information but also about how the process, when done right, can help build capacity within communities, provoke food systems thinking, connect people across scales, and even lead to policy coherence. The authors also highlight how assessment processes make successes and areas for improvement more apparent. Finally, the insights from *Sustainable Food Systems Assessments: Lessons from Global Practice* add more heft to important questions about whose voices are raised through assessment processes and what discourses are reinforced, foregrounded, overlooked, or ignored. In the following sections, we tease apart the challenges and opportunities in

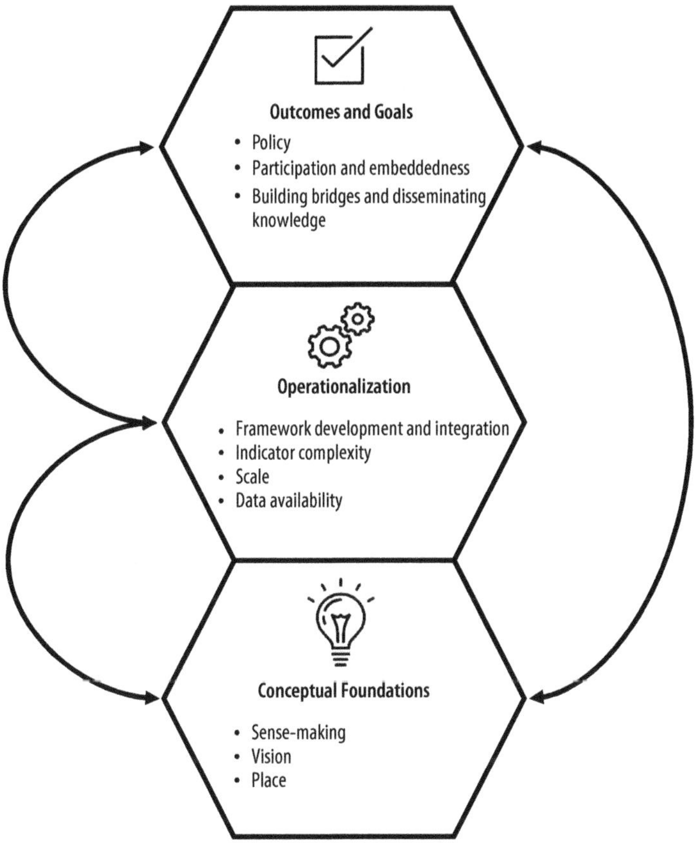

Figure 12.1 Iterative Sustainable Food System Assessment dimensions.
Note: Interaction occurs between and within: Outcomes and Goals, Operationalization, and Conceptual Foundations.

applying these dimensions to sustainable food system assessment (SFSA), drawing heavily from the research in the previous chapters.

Conceptual considerations: sense-making, vision, and place

The conceptual considerations identified during the workshop focused on the process and outcomes of sense-making and vision as well as the need to recognize place-based dimensions (Sonnino et al., 2016). The sense-making used to develop indicators and the lenses used to frame them determine what is included or left out of an analysis. In selecting metrics, Maye and Duncan (2017, p. 268) ask us to broaden 'visibility fields' to question what is visible and why and how this decision-making process unfolds. This also allows us to capture multiple dimensions of performance. It is important to consider 'visibility' as,

> we must be particularly sensitive to aspects which are hidden from our view by the focus on the process of embedding sustainability in the supply chain, and conversely seek to understand how and why our attention is being directed to other areas by the actors concerned and the field of visibility associated with the embedding sustainability in decision-making tool.
>
> (Spence & Rinaldi, 2014, p. 438)

The idea of visibility points to the importance of considering what an indicator represents. As Garnett (2014) points out, using one-dimensional indicators alone, such as efficiency and demand restraint, can be one-sided and oversimplify the context, missing key components that need to be included in sustainability assessment. Frequently, the efficiency perspective promotes the use of technology to meet increasing demand through more production, more efficient chemical use, and waste reduction – reinforcing the technocratic fix and often contributing to a top-down approach. This line of thinking is problematic as it overlooks issues related to food access and quality, including the current reality that there is enough food produced to feed everyone in the world a healthy diet, yet more than 800 million people are food insecure and more than 1 billion people are overweight or obese (FAO, 2017). The second perspective – the demand-driven focus – views solutions as being consumer-led and tackles food challenges from the influence eaters have over the supply chain. In many cases, the focus is on reducing meat-based diets or decreasing food miles to improve both environmental conditions and health. However, this is also too simplistic since indicators that support reduced meat consumption may overlook local sustainable meat production practices that can help sequester GHG and provide a regionally appropriate protein source (D'Silva & Webster, 2017). Alternatively, using a food system transformation lens draws upon the interdependence of production and consumption networks to recognize that sustainable food systems require integrated structural change and that these changes are interrelated and complex (see Chapter 4, this volume). A focus on transformation includes social justice and equity issues and offers a more robust and complete picture of challenges and possible solutions (Garnett, 2014). This integrated vision for fostering a sustainable food system is applied throughout this book. Chapters by Prosperi et al. (Chapter 7), Battersby (Chapter 5), Valette et al. (Chapter 2), Palmer and Santo (Chapter 8), Paredes et al. (Chapter 10), and Cohen (Chapter 11) all make the case for integrating social justice indicators into SFSA analyses. The Spring et al. (Chapter 3) and Meter (Chapter 4) chapters weave social justice into the assessment process itself by elevating the insights of those who had been marginalized as indicators are framed.

As the sustainable food system movement works to transform the industrial food system, it is important to direct the focus away from bottom line, high technology-centred solutions, as well as industry- and export-driven policy to practices focused on social justice, food and nutrition security for all, agroecology, and local circular economies (Anderson, 2019). Food sovereignty is one way to do this as it explicitly values culture, food democracy, the sacredness of food, and the Right to Food (FIAN, 2016; Levkoe & Blay-Palmer, 2018). FIAN's (Food First Information and Access Network) project, People's Monitoring of the Right to Food and Nutrition (RTFN) has the vision that

> Food Sovereignty and RTFN monitoring is consistently used by actors at all levels to result in positive changes in the realization of the RTFN and for the identification of strategic paths to a new society where all human rights are fully realized.
>
> (FIAN, 2016, n.p.)

This project emerged from a desire to collect and report data that includes input from front-line people in civil society organizations (CSOs) and reflects resistance to a technocratic approach. Through this process, food sovereignty and RTFN monitoring information is not only used and reclaimed, but also produced, interpreted, and transformed into action by people and their representatives through different approaches to food, including agroecology, community-based co-operatives, and direct farmer to consumer sales. Food sovereignty and RTFN approaches have been used in a number of contexts, including in recent writing and policy focused on sustainable food system assessment (Levkoe & Blay-Palmer, 2018). While none of the chapters in this volume adopt an explicitly rights-based approach, many contributors to this book advocate for more inclusive processes that work to give agency to marginalized communities (in particular Chapters 3, 4, 5, 6, 8, 10, and 11).

At their best, place-based considerations account for local needs and ground the SFSA process to capture relevant information and point the way to impactful sustainable food system change (Sonnino et al., 2016). For example, by applying an ecosystem-assessment perspective, positing specific foodsheds helps us understand food systems as embedded in broader ecosystems instead of existing exclusively within political boundaries. Such an approach brings an ecosystem viewpoint to the analysis of SFSA based on place through the lens of food (e.g. Kloppenberg, 1996; Mullinix et al., 2016). Place-specific indicators, as well as inclusive and participatory indicator identification and development, are necessary for transformative SFSA processes. Noting the disconnect between indicators at the global and national scale with community initiatives, in particular with needs and goals across scales, Prosperi et al. (2015, drawing from Cassar et al., 2013) observe the importance of developing indicators that include place-based considerations such as the geographical, socio-economic, and cultural context where these tools are implemented. As a result, they stress that, 'a strong and active involvement of the local/community stakeholders is key to design a set of metrics that will be useful to measure real progress and gaps towards the sustainability of the urban food systems' (2015, p. 30). Prosperi's work applying a Delphi method to capture expert insights for the Mediterranean region reported in this book shows that consensus can be used to develop shared indicators (Chapter 7). Meter (Chapter 4) brings this work to a community level, engaging residents in defining linked indicators that cut across issue areas, and express systems levers that can be moved in each context (also City Region Food Systems, Chapter 9).

Place-based assessment can be pivotal for rural and urban authorities who have a chance to facilitate the institutionalization of innovative food system approaches. Research in Marin County, California, makes it clear that using a place-based approach is important. In that case, the analysis focused primarily on food access and food security challenges for low-income households as these factors were identified as having the most relevance in their jurisdiction and were seen as the best levers for change at that time in that place (Marin County, 2012). The URBAL project described in Chapter 2 recognizes the need for both place-based flexibility and robust widespread usability of research tools as it develops an impact-pathway mapping tool to capture sustainable urban food system innovation. Testing this tool in eight cities through 12 innovation labs will help to meet project goals of reach and relevance and enable the maps to capture the place-based vision for each innovation. The Sustainable Cities project in the UK is another excellent example of the importance of considering place throughout the development and implementation of indicators (Chapter 6). The overall goals of this project were: (1) To work with grassroots organizations and local practitioners to define success in cities; and (2) To develop an indicator toolbox to support municipal governments and communities as they work to change the food system. Following a literature review on sustainability and food security indicators, the researchers held four workshops to co-develop a vision with associated metrics across health, economics, and the environment. This information was assembled into an indicator toolbox that was tested in pilot communities. Crossing boundaries, this data is relevant for agencies with respect to the environment, climate change, and economic development. As the project was driven by the needs of the people in each place, there are no standardized objectives or pathways to change since each city is different and so all had different entry points. The Delphi survey used by Prosperi et al. in Chapter 7 developed indicators focused specifically on the Mediterranean region, while Battersby's chapter points to data gaps based on decisions specific to Cape Town.

As part of the work of imagining a sustainable food system, a shared sense of why indicators are being developed, and related goals is important at all steps from analysis and interpretation, through to policy development so that an indicator framework realizes its potential to become a sense-making process. These examples from our book and elsewhere point to ways visions and sense-making as well as place-based considerations enhance the relevance of SFSA. With these visions to guide indicator selection, a related consideration is how to then operationalize indicators. Engaging stakeholders in assessments also transforms the process from place-based sense-making to place-making.

Operationalizing assessment tools

Operationalizing assessment tools relies on several resources including how frameworks are developed and feed into the assessment process, the relative complexity or simplicity of indicators, considerations about scale, and the availability of data.

Frameworks

As discussed in the previous section on conceptual foundations including visioning, a shared framework can capture common goals and lead to a consistent analysis over a broader context by including considerations outside of the narrow project scope (Riley et al., 1999). In addition, the process of sharing/developing common framework(s) can connect and facilitate work towards aspirational goals, such as a common framework with multiple, varied indicators. When selecting metrics, Maye and Duncan (2017) ask us to pay attention to the frameworks we use so that we engage meaningfully with key sustainability challenges (Lakoff, 2010). Morin (2008) recommends a paradigm of complexity that frames more self-reflective assessment work. While Lakoff (2010) refers specifically to the environment, this idea applies equally to enacting sustainable food systems and the crossover with other areas that are, '… intimately tied up with other issue areas: economics, energy, food, health, trade, and security. In these overlap areas, our citizens as well as our leaders, policymakers, and journalists simply lack frames that capture the reality of the situation' (p. 76). At the same time, trends can be identified in the use of assessment tools for more inclusive places. Freedgood et al. (2011) identified several types of community assessments used particularly in the US and Canadian contexts: foodsheds, comprehensive food systems, community food security, food asset mapping, food deserts, land inventories, and food industries. These point to the multi-scale nature of 'place' and how place can be a layered consideration. Frameworks such as the Sustainable Livelihoods Framework (SLF) have been used to ground the work in household and community needs. In Chapter 3, Spring et al. use this framework in a wide range of circumstances including in First Nation communities in the Northwest Territories, Canada, where links were made between climate change, food security, and traditional systems. In particular, political capital was identified as a key community resource. Applying a common framework can enable comparisons across research and community initiatives (Blay-Palmer et al., 2015).

Complexity

There are several challenges that can be considered under the heading of complexity. On the one hand, there is the tension between the need to capture all facets of sustainable food systems dimensions and, at the same time, be simple enough so that indicators are functional. Indicators that are static can misrepresent the bigger picture. The FAO State of Food Insecurity report and the Prevalence of Undernourishment (PoU) indicator referred to in Chapter 1 are cases in point (FAO, 2017). The City Region Food Systems (CRFS) project described in Chapter 9 grappled with these challenges. In that case, researchers looked to develop an approach that was complex and applicable everywhere despite different place-specific issues, capacities, and missions in the pilot cities. The URBAL project (Chapter 2) is aiming to develop a mapping methodology that can be applied across sustainable food system innovations. In these cases, flexibility and multiple options is key so that the approach is relevant across a range of cases. Meter's Chapter 4 also reflects on these SFSA challenges, proposing a complex adaptive systems approach as valuable based on its capacity to reflect actual community needs and levers for systemic change. Complex indicators that address multiple co-benefits are also important in order to reflect complexity through, for example, agro-ecological approaches, so that indicators capture change in more than one dimension and can help to connect knowledge silos.

Scale considerations

It is important to consider how frameworks and indicators can or cannot be translated across scales, ranging from the individual and household to the municipal, regional, and sub-global. Considerations include the way indicators are nested, connected, and/or contradictory. Chapter 1 provides a discussion of the Millennium Development Goals (MDGs), the Sustainable Development Goals (SDGs), and related challenges, including the lack of engagement with community voices, the privileging of scientific knowledge, the disregard for traditional knowledge, and the technocratization of sustainability assessment (Death & Gabay, 2015; Fukuda-Parr & McNeill, 2019). Several authors in this book address these and other challenges related to scale issues. Battersby (Chapter 5) illustrates how data expectations at one scale miss data needs at another scale. The analysis of the food system and food security study commissioned by the City of Cape Town illustrates these disconnects through either the lack of appropriately disaggregated data and/or no data or data requirements by international projects that miss the mark locally. The chapter by Prosperi et al. on the Mediterranean region provides an example of how to address assessment challenges and opportunities at the sub-national level. The chapter describes the consensus-building process developed based on the Delphi survey method to identify and agree on a suite of indicators for food and nutrition security in the context of SFS. The

CRFS project presented in Chapter 9 documents place-specific sustainability dimensions of food flows in the Global South and North. This work relied on either existing or creating multi-scaled, multi-actor networks. These networks were used to enhance urban–rural linkages and improve various dimensions of the food system including food access for low-income families, waste management, improving incomes for rural and urban producers, and by creating food policy (Forster et al., 2015; Dubbeling et al., 2017). As Forster and colleagues explain, '[s]trengthening urban-rural linkages by focusing on improving the holistic performance of food systems at a city region level can contribute to the broader sustainable urbanization agenda. The improvement of city region food systems has significant implications for spatial planning' (Forster et al., 2015, pp. 3–4). Toolkits include indicators that support SFS capacity building and food policy advocacy (Chapters 8 and 9). Paredes et al.'s chapter (Chapter 10) tests indicators at the intersection of sustainable food production and access in three regions in Ecuador. This project emerged from a desire to collect and report data that included and reflected input from front-line people in civil society organizations and demonstrates their resistance to a technocratic approach that would not have represented community priorities. Cohen's chapter (Chapter 11) assesses downstream outcomes around food access in New York City and links them with upstream federal policies around immigration and food, and makes the existing and potential iterative relationships between the two scales clear. Chapter 2 by Valette et al. explores mapping methodologies to support urban and surrounding regional sustainable food system analysis and offers new tools for data analysis. Meter (Chapter 4) discusses the merits of using an adaptive complexity approach within regions as the basis for indicators that are integrative across sectors and scales through both qualitative and quantitative data to highlight core system dynamics. His chapter also informs the city region food-planning approach by identifying the extractive economic structures that place cities in a more powerful position than rural areas. Research focused on assessment at these regional scales can enable more comprehensive and coherent pathways to address food system challenges.

At the city scale, SFSAs typically link food assessment with a range of city goals. As previously discussed, the Milan Urban Food Policy Pact, with its six sustainability categories (production, health and well-being, governance, waste, biodiversity, and social inclusion), provides the basis for urban sustainability assessment and potentially for comparative work across the more than 200 signatory cities. In these cases, comparisons, network, and capacity building can be indirect outcomes of the data collection process. This work is consistent with the chapter by Moragues-Faus (Chapter 6) who assesses the co-production and reflexive processes that occur as indicators are developed during Action Research sustainable food futures projects in the UK.

All of the chapters in this book point to the challenges of interpreting national data at smaller scales as solutions are overly generalized. This points to the relevance of community-based research that leads to the creation of relevant indicators. Based on the work in this book, regional and smaller scales appear to provide the most useful information for policymaking and action, while national and global approaches may add important comparative overarching perspectives through a broader context.

Data availability

Collecting data is expensive and time-consuming and the extent data is or is not available limits and constrains the capacity to support visions. As Battersby (Chapter 5) points out in her chapter, available data is gathered in the context of existing politics. In the case of Cape Town, this has led to data gaps for food system assessment stemming from a misarticulation of data boundaries and the kind of data that was (or was not) available. Pointing to the disconnect between food and landscapes, Spring et al. (Chapter 3) discuss the lack of data covering remote northern communities and the gaps that need to be filled to link traditional food systems with the health of the land as a key source of regional food and boreal ecosystem health. The question of data availability is critical in communities closely tied to the land for their food as the climate continues to change. Paredes et al. (Chapter 10), Moragues-Faus (Chapter 6), Valette et al. (Chapter 2), and Palmer and Santo (Chapter 8) also address the question of data availability, in these cases through participatory processes as part of the development for SFSA.

When data is available or is being collected for the first time, it is important to be able to compare indicators over time and build on that work to keep it relevant, identify trends, and share it out as well as identifying data gaps. The Food Counts report card for Canada is a good example of this challenge. As a key part of the analysis, Levkoe & Blay-Palmer (2018) identified the need for new metrics. For example, there was no existing way to measure 'food is sacred' as one facet of the Canadian SFSA. Identifying data needs emerged as an important contribution of that work. In Maryland, an intuitive, accessible metric was developed to capture the siltation of a river (Flora, 2018). The goal was to make the water quality problem visible to local residents who lived upstream so they would be motivated to make a difference. To make the problem clear, the mayor put on white tennis shoes and walked out into the bay to see where he lost sight of his feet, recalling that people used to be up to their necks and still see their feet. Every year at the same time he would walk to the point where he could still see his feet and this measure would be posted on a bulletin board. As the siltation decreased over time, in part though work with farmers up and down the watershed, people celebrated enthusiastically as results were reported widely throughout the state. This was a very simple, inexpensive but visible indicator of turbidity that could be linked directly to soil run-off, making the remediation challenge and the progress clear for local communities.

Outcomes and goals

Outcomes and goals for sustainable food system assessment are clearly context dependent. In many cases they include policy generation, which is fostered by, and helps to enhance, the engagement and inclusion of participants and communities, builds connections between initiatives, embeds change in communities and allows for widespread knowledge dissemination.

Policy generation

Several questions emerged from the Toronto workshop around policy as an outcome of data gathering based on research projects. These questions focused on: (1) The conditions needed for policy to emerge from SFSA work; and (2) Whether there is a common process or trajectory that all/most places follow as they work towards SFS. Key informants at the workshops also identified power dynamics on how/whether policy is implemented as an important consideration. The extent to which subsidiarity applies is also key as it reflects the extent of empowerment of local actors and the extent to which power and agency are devolved as close to the community scale as possible. The IPES-Food 2017 report provides insights into necessary conditions for policy formulation at the city scale. Drawing on in-depth interviews and literature scans, the report identifies the policies and conditions needed to create and maintain sustainable food systems in cities including: (1) The need for robust inter-sector actor networks as channels for policy influence and the basis for partnerships; (2) The importance of partnerships between municipal departments and external organizations to allow for co-governance. This requires supportive resources and capacity for implementation; (3) The determination of formal governance and terms of reference so all actors know what is expected and are accountable; (4) Conducting research and monitoring impacts to demonstrate efficacy, and to identify and remedy unexpected impacts; and (5) Focusing on areas of local government control and influence seeking synergies with the national level where possible (IPES-Food, 2017).

The SFSA processes described in *Sustainable Food Systems Assessment: Global Approaches to Practice* support and build from the insights captured in the IPES findings to help address food systems challenges. In some of the research reported in this book, conflicts emerge between different levels of government (Chapter 11). In other cases, data is mis/realigned across departments, sectors, or scales, as in the findings reported by Battersby (Chapter 5) and Moragues-Faus (Chapter 6). As reported, indicators can help deal with these tensions as they can enable the identification of which policy levels sh(c)ould be addressed – in some cases national policies are needed to provide an overarching context, in other cases local zoning may need to be changed (Chapter 8). This type of policy alignment is ongoing as circumstances evolve. A related set of questions revolves around whether the assessment uses universal indicators or not. And, whether researchers want to identify indicators that will help decide which policy level is the best for intervention. To help answer these questions, there is an ongoing exploration about whether international initiatives can be leveraged to shift from technical to more policy approaches. For example, in their policy-directed assessment, Perez-Escamilla et al. (2017) evaluate the usefulness of national and global food security indicators to make the case for indicators that facilitate evidence-based policymaking. Using SMART (specific, measurable, achievable/attainable, relevant, and time-bound) criteria they assess the usefulness of commonly used indicators (Dietary Diversity Measures; Measures Based on Participatory Adaptation; Prevalence of Undernourishment; Global Hunger Index; Global Food Security Index; Dietary Record; 24-hour recall; Food Frequency Questionnaires; Food Consumption Score; Household Dietary Diversity Score; Coping Strategies Index; Experience-based food insecurity scales; Anthropometry) for policymakers and classify indicators for usability. Battersby's chapter raises these important points from a community and local perspective providing insights into how indicators are shaped by the existing hegemony and where there may be entry points for change (Chapter 5). In Chapter 8, Palmer and Santo offer insights into the advocacy process through their work on food policy councils. Their extensive survey of existing food policy initiatives points to ways that advocacy may be improved. And, as Prosperi et al. reiterate, the ways in which assessment results get translated, packaged, and mobilized are important factors in how change occurs (or does not occur) in food systems at all levels (Chapter 7). If policymakers do not understand interactions and dynamics that are inherent to SFS then assessment results can be misinterpreted and policy and programmes will be ineffective or even counter-productive to the goals policymakers and communities establish (Chapter 11).

Participatory approaches and embeddedness

When working in communities, comparing and talking about developing or replicating a process can help to identify relevant indicators, build networks, and embed learnings. All of the chapters in this book support the importance of communities as they need to determine their own values and related visions and what they want to work towards based on their goals, needs, and resources. As discussed, international project goals can be set at the global scale with indicators identified for specific national or local concerns. The SDGs are an interesting example of this approach; while there was consultation with member states, intergovernmental organizations, non-governmental organizations, business sector, and other major groups, there was very little public consultation. Some of these concerns include the need for continuity between the process that established the SDGs and their implementation (Palmer, 2015), issues around land rights (Wise, 2015), and the technocratization of the process to the exclusion of smallholder farmers and others (IPES-Food, 2015). Participatory research by Sanye-Mengual et al.

(2018) in Bologna, Italy, into the interconnections between local urban food production (UFP) and global sustainability initiatives, in particular the SDGs, identified the need to consider sustainability transversally and through a participatory approach. Enhancing the chance of success and uptake,

> [t]his bottom-up approach unveiled a comprehensive vision of sustainable UFP, the relevance of certain sustainability elements and key aspects to take into consideration for the implementation of UFP, the design of effective policy-making and the development of research studies on the sustainability of UFP that built upon the presented conceptual framework.
> (Sanye-Mengual et al., 2018, p. 15)

This building process is important to embed indicator work so that it has staying power and is not lost with a shift in government or through other changes. One mechanism is to liaise with official data collection agencies as this can help to maintain and/or add indicators to the data-gathering process and embed food system sustainability. This is also important to ensure the continuation of functions needed to support food systems.

As explored earlier, creating coherence across scales for indicators is challenging – community participation in developing action-oriented assessment can help mitigate this problem. For example, in Montpellier, France, university researchers gathered municipally elected people together to ask them what they were doing in their political work to advocate for their own food systems. While food was not previously a focus, the elected officials involved in the project started to talk about school canteens, land use, environment, shops, and then the foodscape. Then suddenly everybody did work for food. It was the creation of the process that generated a sense of food, the potential of food, and suddenly people could feel that working on food could be of benefit. Then, when food was an issue and they were working on that, they realized that they need data to understand the situation, to know, for example, that people are hungry, to map the situation, then data becomes important. The data gathering also allows people in civil society and policymakers to think beyond the agenda they had already established. This was the case based on the research in South Africa as reported by Battersby in Chapter 5. Meter found in Chapter 4 that simple network maps changed economic development leaders' perspectives. Prosperi et al.'s chapter (Chapter 7) used a participatory research method throughout the indicator development process to ensure relevant co-production of knowledge and to facilitate, given the complexity, that everyone understood terms, concepts, and frameworks in the same way (see also Lehtonen et al., 2016). Palmer and Santo (Chapter 8) also demonstrate the importance of iteration as assessment tools are being developed as well as the need for as many face-to-face consultations as possible to enable this iteration. The URBAL project (Chapter 2) and the CRFS work (Chapter 9) demonstrate the possibility of engaging multiple stakeholders across disciplines, sectors, and scales and the benefits that can accrue, including increased capacity and coherence for communities.

Building bridges and disseminating knowledge

Part of thoughtful indicator creation is ensuring outcomes do not reinforce or seed more silos between disciplines, sectors, and government departments and institutions. Rather, indicators need to allow actors to talk to each other more frequently and more effectively. At the same time, it is important to work from a food systems perspective to provide the resources and data that communities need in order to address the issues they value within their own food systems. For example, at a meeting leading up to the 3rd UN Conference on Housing and Sustainable Urban Development (Habitat III, Quito, 2016), a panel whose diverse participants worked at multiple scales and from many institutions discussed the need to share instruments and knowledge among cities. While they identified the need for instruments to carry out assessments as well as evidence from focused studies, planning and regulatory instruments for cities were singled out as lacking. The co-creation of knowledge provides a more inclusive and relevant set of indicators as the basis for policy and policy tools.

In this context, a question becomes, how can indicators be used to create networks that link across sectors and institutions to foster system transformation? The indicator work needs to be embedded in potential policy delivery, which is not always easy. The cities of Toronto, Canada and Milan, Italy are pioneers in this regard having developed both depth and breadth in food policy. The Toronto Food Policy Council is a world leader having been established nearly 30 years ago, while Milan has become extremely active following the launch of the Milan Urban Food Policy Pact in 2015. To continue this work requires unpacking the 'policy-governance' box and looking at how food can become an issue that engages all communities. Questions to address include: (1) is policy integration taking place, for example between planning and land use (Forster et al., 2015)? And (2) if it is, is the balance right? The CRFS process reported in Chapter 10 and the idea of linking upstream policy with downstream impacts as proposed in Chapter 11 by Cohen are good examples of how this can work.

Sharing the results from indicator projects is key and should be considered part of the process from project inception. There is a need to actively share findings with other researchers, government, the private sector, and civil society. Using multiple platforms including clear plain language reports, online spaces, and social media is critical to keep forward momentum so projects are not time bound and the work is taken up and used by administrators and policymakers. Creative partnerships could help ensure findings are accepted and acted upon, and also employed as part of public education. Including key actors from the outset also helps with knowledge dissemination, as indicators are more likely to reflect key player needs and goals.

It is also valuable to look outside food systems to learn from other domains, such as the health sector where a lot of relevant work is already being done, for example, the World Health Survey and the water–energy–food nexus. Sustainability movements are confronting similar problems as they challenge the neo-liberal hegemony, so it is helpful to share information about processes and the kinds of frameworks that are useful. The UN Convention on Biodiversity (CDB) was among the first, multilateral agreement to tackle a global challenge with the recognition of 'traditional knowledge' and provides an example of integration across multi-stakeholders.

The view from here

We conclude the book with some insights that pull out and weave together some common threads as well as raise questions to consider moving forward as we work to advance the relevance and breadth of sustainable food system assessments.

First, in addition to data gathering and tool development, an important part of the assessment work at the centre of this book is the development of, giving voice to, or reinforcing socio-political processes founded in social movements and the evolving relationship between policymakers, practitioners, civil society, and academics. Working on assessments can have the added benefit of capacity building through the participation enabled with this type of research (Chapter 9).

It also raises questions about what constitutes data. While indicators are important, stories are interesting and compelling; thus, a key question is, how do we capture the most useful insights? And, how do we link and talk about the indicators as nested, linked, *and* relational? It is important to talk about processes and purpose simultaneously. Chapter 3 by Spring et al. is an excellent example of how communities with traditional food systems need to connect their community food system assessments to the health of the land. In this case, the well-being of the boreal forest, including caribou and fish health, is part of understanding the sustainability of Kakisa's food system. The EKOMER project (Chapter 10) focuses attention on the household as part of a city region while chapter authors Paredes et al. (Chapter 10) explain the power of the,

> efforts of consumers to self-organize around ethical values and morals of consumption and to exert political influence at any stage of the process. Such is the notion of 'co-producer', 'a consumer who knows and understands problems of food production' (Petrini, n.d.)
>
> (p. XX)

Colombo used a city region focus as part of the FAO-RUAF CRFS project (Chapter 9). In that case, stakeholders attended primarily to indicators using their locally determined project foci of waste, food security, or food safety. The food policy work by Palmer and Santo (Chapter 8) demonstrates how assessment can be used as a vehicle for building tools for activism. It is anticipated that the URBAL project (Chapter 2) will build communities of food around the process of mapping sustainable urban food system innovations so that knowledge is co-created and agency is enhanced as networks grow. As part of these processes, it is important to combine and point out how the indicators are linked and improve the connections as learning opportunities to understand more about food systems as a whole. In considering scale of assessments, questions include how/whether to move from local or regional indicators to more national or international indicators, or the reverse, and what links and connections exist between scales? Battersby's, Meter's and Cohen's work (Chapters 4, 5 and 11) helps us to understand the challenges in shifting between scales and the need to work within complex frameworks. All of the projects in this book point to the centrality of iteration and the value of cross-verifying data as it is gathered.

Second, while the initiatives reported in this book help to broaden the conversation about assessment and policy, more is needed. In addition to understanding local food systems, it is also important to understand how international laws affect local laws, for example the World Trade Organization, the European Union, or through the multi- or bi-lateral accords such as the SDGs. The common list of indicators used by international organizations applies a global lens that can be difficult to apply at the local level (Chapter 5). It is important to have standard indicators but, from a governance point of view, if people do not understand the value of food then it is a challenge to integrate food into policies. Moving to standard indicators coupled with a participatory approach for interpretation and implementation can be effective. Related questions to address include, does future work compare processes, frameworks, and/or specific indicators? It seems there is an interesting nexus between indicators and unknowns in the realm of governance that needs to be further unpacked to try to identify ways to make new pathways between different interests. A related question is what are the indicators we can use to create and signal transformative food systems? One way to know the indicators are successful, is that we would see change in the values of indicators over time, or pertinent, new indicators may be initiated. But what combination of indicators would signal that food system sustainability is being achieved? The authors of this book certainly suggest that indicators of social capital and relationships of trust play a central role, in concert with other measures (for example, the role of communities as discussed in Chapters 2, 3, 4, 9, & 10, this volume).

A third consideration is capturing the invisible dimensions of SFSA. The social economy, particularly as it is articulated in informal economies, is a good example of the largely invisible and often poorly measured dimensions of SFSA, a situation that is captured in the work of Joubert et al. (2018) on the ways in which informal food systems in South Africa, Kenya, Zimbabwe, and Zambia are overshadowed. Case studies or other qualitative processes that capture these important stories can complement metrics and other assessment tools by providing the depth needed to connect the data in a meaningful way. To develop relevant policy, it is critical to capture the practices that are out of the scope, or beyond the 'official' framework, or simply invisible and the associated contributions that are very difficult to measure. There are many data challenges including, as reported by Paredes et al. (Chapter 10), with people being reluctant to be 'data-ed'.

The book also raises questions about complexity and using an adaptive systems approach. Meter (Chapter 4) is the most clear about this opportunity but others, for example the chapters by Valette et al. (Chapter 2), Cohen (Chapter 11), and Santini et al. (Chapter 9), draw on these assumptions. As Kate Clancy (2014) states, complex adaptive systems integrate and rely on, 'many diverse and autonomous components or parts ... which are interrelated, interdependent, linked through many (dense) interconnections, and behave as a unified whole in learning from experience and in adjusting (not just reacting) to changes in the environment' (p. 10). The multi-directional, iterative co-creation of knowledge demonstrated in the development of the advocacy toolkit developed by Palmer and Santo (Chapter 8) is an excellent example where multiple individuals in the same organization completed the assessment as they worked towards enriched, varied convergence. The heterogeneity that comes from multi-stakeholder, multi-sector, and multidisciplinary assessment opens up space for verification and buy-in.

Finally, we need to consider enacting assessment processes and tools to help address the pressing issues the world now faces. Climate change, migration shifts, and growing inequality can all be addressed through more sustainable food systems. To do this, we need assessment processes and tools that reflect the realities of those most disadvantaged in households, communities, regions, and countries. By providing relevant assessment support, we can make decisions based on evidence that raises the voices of the people who need to be heard the most. At its best, a sustainable food system assessment process, as demonstrated by all the chapters in this volume, offers the potential to build capacity and bring transparency and clarity, in turn enabling a better use of resources and learning over time and across scales (Anderson, 2015). They can also provide the basis for seeing how participation may need to change, measure change over time, enable strategy development, knowledge transfer, and inform transformative, coherent policy.

Acknowledgements

We would like to acknowledge the support of the Social Sciences and Research Council of Canada Partnership Grant, and the Laurier Centre for Sustainable Food Systems. We are also grateful to our community partners for their ongoing engagement and support. Square check mark icon in figure made by Freepik from www.flaticon.com. Light Bulb icon in figure made by Puppets from www.flaticon.com. Gear icon in figure made by Freepik from www.flaticon.com.

References

Anderson, M.D. (2015) The role of knowledge in building food security resilience across food system domains. *Journal of Environmental Studies and Sciences*. 5(4), pp. 543–559.

Anderson, M.D. (2019) Comparing the effectiveness of structures for addressing hunger and food insecurity. In: Andrée, P., Clark, J.K., Levkoe, C.Z., & Lowitt, K. (Eds.) *Civil society and social movements in food systems governance*. London, Routledge, pp. 124–144. doi:https://doi.org/10.4324/9780429503597.

Blay-Palmer, A., Renting, H., & Dubbeling, M. (2015) City region food systems: A literature review. *Carasso Foundation*. Report available at: https://www.ruaf.org/sites/default/files/City%20Region%20Food%20Systems%20literature%20review.pdf.

Clancy, K. (2014) A different way to approach policy change. *Journal of Agriculture, Food Systems, and Community Development*. 4(4), pp. 9–11.

Death, C. & Gabay, C. (2015) Doing biopolitics differently? Radical potential in the post-2015 MDG and SDG debates. *Globalizations*. 12(4), pp. 597–612.

D'Silva, J. & Webster, J. (Eds.) (2017) *The meat crisis: Developing more sustainable and ethical production and consumption*. London, Routledge.

Dubbeling, M., Santini, G., Renting, H., Taguchi, M., Lançon, L., Zuluaga, J., De Paoli, L., Rodriguez, A., & Andino, V. (2017) Assessing and planning sustainable city region food systems: Insights from two Latin American cities. *Sustainability*. 9(1455). doi:10.3390/su9081455.

FAO (2017) *State of food security and nutrition in the world*. Rome, FAO. Available at: http://www.fao.org/3/a-i7695e.pdf.

FIAN (2016) *FIAN International Annual Report 2016*. Available at: https://www.fian.org/fileadmin/media/Publications/Annual_Report/FIAN_International_Annual_Report_2016_english_march.pdf.

Flora, C.B. (2018) *Rural communities: Legacy + change*. London, Routledge.

Forster, T., Santini, G., Edwards, D., Flanagan, K., & Taguchi, M. (2015) *Strengthening urban rural linkages through city region food systems*. UNCRD/UN Habitat issue of Regional Development Dialogue.

Freedgood, J., Pierce-Quiñonez, M., & Meter, K. (2011) Emerging assessment tools to inform food system planning. *Journal of Agriculture, Food Systems, and Community Development*. 2(1), pp. 83–104.

Fukuda-Parr, S., & McNeill, D. (2019) Knowledge and politics in setting and measuring the SDGs: Introduction to special issue. *Global Policy*. 10, pp. 5–15.

Garnett, T. (2014) Three perspectives on sustainable food security: Efficiency, demand restraint, food system transformation. What role for life cycle assessment? *Journal of Cleaner Production*. 73, pp. 10–18.

IPES-Food (2015) *The new science of sustainable food systems: Overcoming barriers to food systems reform*. First Report of the International Panel of Experts on Sustainable Food Systems.

IPES-Food (2017) *Too big to feed: Exploring the impacts of mega-mergers, consolidation and concentration of power in the agri-food sector*. Available at: http://www.ipes-food.org/_img/upload/files/Concentration_FullReport.pdf.

Joubert, L., Battersby, J., & Watson, V. 2018. *Tomatores and taxi ranks: Running our cities to fill the food gap*. Cape Town, South Africa, African Centre for Cities, University of Cape Town.

Kloppenberg, J., Hendrickson, J., & Stevenson, J. W. (1996) Coming into the foodshed. *Agriculture and Human Values*. 13(3), pp. 33–41.

Lakoff, G. (2010) Why it matters how we frame the environment. *Environmental Communication*. 4(1), pp. 70–81.

Lehtonen, M., Sébastien, L., & Bauler, T. (2016) The multiple roles of sustainability indicators in informational governance: Between intended use and unanticipated influence. *Current Opinion in Environmental Sustainability*. 18, pp. 1–9.

Levkoe, C.Z., & Blay-Palmer, A. (2018) Food counts: Food systems report cards, food sovereignty and the politics of indicators. *Canadian Food Studies/La Revue canadienne des études sur l'alimentation*. 5(3), pp. 49–75.

Marin County (2012) *Marin community food systems assessment project report: Fixing food together.* Available at: https://marinfoodsystem.files.wordpress.com/2012/02/marin-county-food-system-assessement-project-report.pdf.

Maye, D. & Duncan, J. (2017) Understanding sustainable food system transitions: Practice, assessment and governance. *Sociologia Ruralis*. 57(3), pp. 267–273.

Morin, E. (2008) *On complexity*. Cresskill, NJ, Hampton Press.

Mullinix, K., Dorward, C., Sussmann, C., Polasub, W., Smukler, S., Chiu, C., Rallings, A., Feeney, C., & Kissinger, M. (2016) *The future of our food system: Report on the southwest BC bioregion food system design project*. Richmond, British Columbia, Institute for Sustainable Food Systems, Kwantlen Polytechnic University.

Palmer, E. (2015) Introduction: The 2030 agenda. *Journal of Global Ethics*. 11(3), pp. 262–269.

Perez-Escamilla, R., Gubert, M.B., Rogers, B., & Hromi-Fiedler, A. (2017) Food security measurement and governance: Assessment of the usefulness of diverse food insecurity indicators for policy makers. *Global Food Security*. 14, pp. 96–104.

Prosperi, P., Moragues-Faus, A., Sonnino, R., & Devereux, C. (2015) *Measuring progress towards sustainable food cities: Sustainability and food security indicators*. Report of the ESRC Financed Project 'Enhancing the Impact of Sustainable Urban Food Strategies'. Swindon, ESRC.

Riley, F., Mock, N., Cogill, B., Bailey, L., & Kenefick, E. (1999) *Food security indicators and framework for use in the monitoring and evaluation of food aid programs*. Washington, DC, Nutrition Technical Assistance Project (FANTA).

Sanyé-Mengual, E., Orsini, F., & Gianquinto, G. (2018) Revisiting the sustainability concept of urban food production from a stakeholders' perspective. *Sustainability*. 10(7), p. 2175.

Sonnino, R., Marsden, T., & Moragues-Faus, A. (2016) Relationalities and convergences in food security narratives: Towards a place-based approach. *Transactions of the Institute of British Geographers*. 41(4), pp. 477–489.

Spence, L.J. & Rinaldi, L. (2014) Governmentality in accounting and accountability: A case study of embedding sustainability in a supply chain. *Accounting, Organizations and Society*. 39(6), pp. 433–452.

Wise, T.A. (2015) *The great land giveaway in Mozambique*. Dollars & Sense. Online. March/April. Available at: http://dollarsandsense.org/archives/2015/0315wise.html.

Index

Note: page numbers in *italic* type refer to Figures; those in **bold** type refer to Tables.

For Product Safety Concerns and Information please contact our EU representative GPSR@taylorandfrancis.com
Taylor & Francis Verlag GmbH, Kaufingerstraße 24, 80331 München, Germany

www.ingramcontent.com/pod-product-compliance
Ingram Content Group UK Ltd.
Pitfield, Milton Keynes, MK11 3LW, UK
UKHW010957180926
475701UK00004B/414
* 9 7 8 1 0 3 2 0 8 3 9 3 3 *